# 100 SCIENCE DISCOVERIES

that changed the world

Published in the United Kingdom in 2021 by
Pavilion Books
43 Great Ormond Street
London, WC1N 3HZ

Produced by Salamander Books, an imprint of Pavilion Books Group Limited.

Photo Credits: Alamy - Sipa US, Zoonar GmbH, NASA, ESA, Lebrecht Music & Arts, Photo 12, The History Collection, Science History Images, REUTERS, North Wind Picture Archives, Universal Images Group North America LLC, Marmaduke St. John, Jo Ingate, Theodore Liasi, World History Archive, Laslo Szirtesi, AF Fotografie, Olivier le Moai, Angelo Homak, Agefotostock, Album, Jason Finn, Martin Bond, Stuart Fretwell, Voigt Archive, Stanislav Sergeev, Bradley White, The Print Collector, Pictorial Press, Robert Matton AB, INTERFOTO, Prisma Archivo, Natural History Museum, Susan E. Degginger, Tetra Images, Granger Historical Picture Archive, Allan Hartley, Classic Image, Irina Dmitrienko, Utcon Collection, DPA, Sueddeutsche Zeitung, Steve Taylor ARPS, nobeastsofierce Science, World of Triss, MCS Illustrations, Nigel Cattlin, HS Photos, Joe Belanger, BSIP SA, Molekuul.be, Zvonimir Atletic, Oliver Hoffman, Juriaan Wossink, Yuri Moroz, Michelmond, Andrea Danti, Quantstock, Artsiom Petrushenka, Archivo GBB, Neko, Cynthia Lee, Jupe, ClassicStock, imageBROKER, Everett Collection Historical, Olha Rohulya, StockTrek Images, DOD Photo, Ikonacolor, IanDagnall Computing, Incamerastock, RBM Vintage Images, Dinodia Photos, Shyamala Muralinath, Alex MacNaughton, Michael Piepgras, PhotoEdit, Graham Clark, David Levenson, Keystone Press, United Archives GmbH, PA Images, Stanislav, PhotoSpirit, Pascal Boegli; Library of Congress; B.T. Batsford Archive, Pavilion Image Library, Polly Matzinger.

10 9 8 7 6 5 4 3 2 1

ISBN: 978-1-911663-54-6

Printed in China

# 100 SCIENCE DISCOVERIES

## that changed the world

—

Colin Salter

PAVILION

# Contents

*RIGHT: The Hubble Space Telescope, with its lack of background light or atmospheric interference, has allowed scientists to look deep into space from 1990, with a projected end of mission around 2030.*

# Introduction

The history of science is measured in milestones of discovery. Each new milestone allows other scientists to further advance the sum of human knowledge. As Sir Isaac Newton said, "If I have seen further, it is by standing on the shoulders of giants."

American sci-fi author Frank Herbert noted that "the beginning of knowledge is the discovery of something we do not understand." It is a human condition to make sense of our surroundings. Homo sapiens is an inquisitive species and it is a scientist's profound curiosity that brings discoveries, pushing at the boundaries of the known world and bringing order to chaos.

A distinction should be made early on between discovery and invention. There is a significant difference between discovering a rule or a law of pure science and

the harnessing of that knowledge to transform the world. No better example of this can be found than Heinrich Hertz, who established the existence of radio waves, but could think of no real use for them. Invention is the employment of discoveries, and that's the subject for another book. This one hand-picks one hundred of the most significant discoveries in history, made by the sharpest, most curious minds in science.

History tends to remember discoverers rather than inventors. That is perhaps unfair to inventors; but if asked to name some famous scientists the average person would probably come up with those who made breakthroughs in knowledge: Marie Curie, who discovered radium; Isaac Newton and his Laws of Motion; Alexander Fleming who revealed the properties of penicillin; not forgetting Einstein, Galileo, Pasteur and the rest.

They are all here, starting with Euclid, whose codification of geometry was man's first attempt to quantify his environment – the very word means "land measurement." The urge to bring order out of chaos through measurement and classification is the beating heart of science. It might be a list of the elements, a knowledge of different blood groups, or the melting points of superconductors; but when someone wanted to know why some materials burn, why the body rejects some blood transfusions, or what happens when you freeze certain gases, they were taking the first steps in discovering something.

*LEFT: Arthur Eddington's capture of images from the 1919 total eclipse advanced many areas of science.*

## HAPPY ACCIDENTS

Not all great scientists find what they are looking for; but the best are able to identify an anomaly and pursue it. Alexander Fleming, for example, had been studying bacteria before his annual holiday. He cleared his Petri dishes to one side before he left so that his colleague had space to work. On his return two weeks later he found that fungus, borne on the air from the room below his laboratory, had contaminated one of the dishes and killed off the bacteria which he had been growing.

The fungus was penicillin, the first great antibiotic, and Fleming was always modest about his fame as its discoverer – "the Fleming effect" as he called it. It's true that the whole thing was an accident caused by an open window and someone else's dirty room. But more than one scientist has addressed the question of accidental discovery. "A discovery," said Albert Szent-Gyorgyi, the Hungarian biochemist who first isolated Vitamin C, "is an accident meeting a prepared mind."

*BELOW: Some of Louis Pasteur's original equipment on display at the Musée Pasteur, part of the Pasteur Institute in Paris.*

"Accidents never happen accidently," said Andre Geim of Manchester University, who along with Konstantin Novoselov discovered the properties of wonder material graphene in 2004. "Good scientists create the environment for as many as possible of those accidents to happen." In other words, a discoverer has to be in the right place with the right frame of mind to learn from the right mistakes.

## STANDING ON THE SHOULDERS OF GIANTS

Hundreds of millions of lives have now been saved with penicillin and other antibiotics, thanks to Fleming's "accident". But that would not have been possible without Antonie van Leeuwenhoek, a seventeenth-century Dutch draper, whose fascination with microscopes led him to discover bacteria. And as genetics pioneer Sir Henry Harris said, "Without Fleming, no Chain *[who discovered the molecular properties of penicillin]*; without Chain, no Florey *[who conducted the first treatment with the antibiotic]*; without Florey, no Heatley *[who discovered a way of mass-producing it]*; without Heatley, no penicillin."

ABOVE: *His name has become synonymous with genius, Albert Einstein in 1921. However, it is comforting to note he didn't get everything right, and disputed the Big Bang theory.*

BELOW: *Wilhelm Röntgen was experimenting with a Crookes (electrical discharge) Tube when he stumbled across the phenomenon of X-rays.*

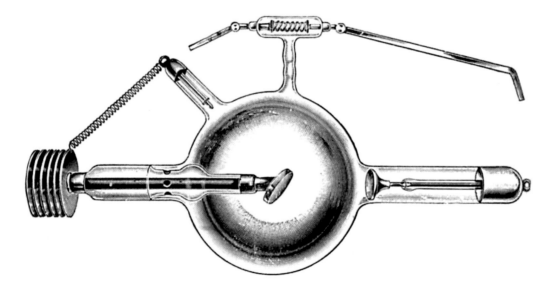

Scientists depend on their predecessors' discoveries to make discoveries of their own. Another science fiction author, Isaac Asimov, observed that, "there is not a discovery in science, however revolutionary, however sparkling with insight, that does not arise out of what went before." Scientists have always depended on the groundwork laid down by their forebears. Even that most baffling of physical arenas, quantum theory, was the result of a set of scientific laws discovered over two millennia which worked for earthbound events but could not fully explain all cosmological phenomena.

In an increasingly complex scientific world, scientists also depend more and more on their contemporaries, either working together or corroborating each others' discoveries. Gone are the days when enthusiastic amateurs such as van Leeuwenhoek, with a microscope in the back of the shop, could make paradigm-shifting observations. Science began to receive professional status toward the end of the eighteenth century – William Herschel, who discovered the planet Uranus, received a stipend when he was appointed Astronomer Royal in the court of British king George III; and his sister Caroline Herschel, who discovered many comets, was the first professional female scientist in 1787, when she became his paid assistant.

*ABOVE: Marie Curie working with daughter Irène Joliot-Curie, who, along with her husband Frédéric, won the 1935 Nobel chemistry prize for the discovery of artificial radioactivity. The Curies' success highlights the folly of restricting education and roles for women in science for hundreds of years.*

## DISCOVERIES IN THE TWENTY-FIRST CENTURY

Today, discoveries are most likely to be made by teams of scientists. Funding for the places where their "accidents" may occur run into many millions and so do the commercial returns expected from them. Like all other areas of human endeavour, science is now a global and often corporate effort. Never has this been more apparent, or more laudable, than in the recent urgent search for a vaccine against the Covid-19 virus which has taken such a toll of the world's population in 2020 and 2021. Although Big Pharma hopes to reap the rewards, scientists made the discoveries. And humanity is the real winner.

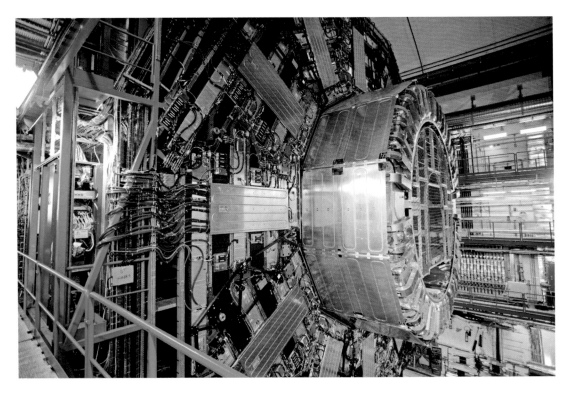

*ABOVE: The Compact Muon Solenoid detector attached to the Large Hadron Collider at CERN in Switzerland. It was built to study the outcomes of proton-proton collisions.*

The history of scientific discovery continues to take giant leaps. During the writing of this book, researchers at CERN's Large Hadron Collider near Geneva (where the Higgs boson was finally identified) have discovered non-conforming behaviour in one kind of particle (the B-meson), while others at Fermilab in the US have observed abnormal responses from another (the muon) under certain conditions. While the world is still adjusting to quantum theory, these two observations may herald a completely new approach to the science of particle physics.

Elsewhere it was announced that Artificial Intelligence (AI) had solved a problem which had stumped biology researchers for fifty years. AlphaFold, an AI programme by the celebrated DeepMind laboratory in London, has discovered how proteins fold themselves into complex three-dimensional shapes.

This mechanism is central to the processes of life and understanding it has benefits for nutrition and medicine, healing and feeding the world, and even ridding it – via green enzymes – of some of mankind's pollution of it.

WHO GETS THE CREDIT?

Of course AlphaFold couldn't have done it without Frederick Sanger who discovered that proteins are chains of amino acids; and Sanger built on the insights of Linus Pauling, who predicted the secondary structures of proteins; and Pauling developed the ideas of William Cumming Rose who continued work begun by Thomas Burr Osborne, who owed much to Dutch chemist Gerardus Johannes Mulder who first described proteins.

But Mulder was describing things which had first been proposed by French chemist Antoine Fourcroy. So who should get the credit, from Fourcroy to AlphaFold?

To return to penicillin, Fleming will forever be credited with its discovery. But the use of fungus in healing has been part of folk medicine for hundreds if not thousands of years. In the late nineteenth century Arab stable boys were known to apply mould to the sores on horses' legs; and Ancient Egyptian medicine men are known to have applied fungi and flora to infections.

After millennia of discoveries, much of the chaos of the world has been made sense of by scientists. Today, it is the least accessible areas where scientists are making new discoveries, either in distant galaxies, on nearby planets or at the level of subatomic particles. There is still much to find out in the depths of our oceans and in the complexities of the human brain, and if we are to survive on this planet, in the capture of what Joseph Black once described as "fixed air" – carbon dioxide.

*BELOW: NASA's Perseverance Mars Rover and the Ingenuity Helicopter on the Martian surface. The Perseverance successfully landed on February 18, 2021 to begin the search for signs of ancient microbial life.*

# Euclid

## (c.323–283 BCE)

## Geometry

He might not have had a word for it, but prehistoric man was already using forms of geometry to mark out the landscape and to track the movement of the heavens. In the third century BCE one Greek man's grasp of geometric rules defined the science.

Geometry concerns the shape of things – their angles, their lines, curved and straight, and the relationship between them. It is used to calculate distance, area and volume, the three dimensions which define our world. The roots of this mathematical science lie in the instinctive desire to make sense of the environment. The word "geometry" comes from the Greek words for "land" and "measurement" and alignments of prehistoric structures in the landscape are one example of our ancestors' use of geometry.

By the third century BCE mathematicians in the Indian subcontinent and the Greek peninsula were developing advanced geometric ideas. One man, Euclid of Alexandria, pulled all these ideas together and unified them under their new name – Geometry. Alexandria in Egypt was a new city, founded less than a hundred years earlier by Alexander the Great, one of many cities to be given his name. It became a centre of learning, famous for its library (the largest in the world at the time) and for its architecture: the Lighthouse of Alexandria was one of the Seven Wonders of the Ancient World. Euclid was a product of this enlightened metropolis.

Euclid published the sum of his mathematical knowledge in a series of thirteen books called *The Elements*, in around 300 BCE. Copies made by professional scribes circulated his ideas throughout the known world. The oldest surviving handwritten copy was made in around 900 CE.

The first six volumes of *The Elements* deal with two-dimensional geometry – the geometry of triangles, rectangles, circles and polygons, angles and proportions, and the construction of the Golden Ratio. The next four concern Number Theory – prime numbers, perfect numbers, sequences, highest common factor and lowest common denominator. The final three return to geometry and progress from two to three dimensions, looking at cones, pyramids, cylinders and platonic solids – three-dimensional figures with sides made of equal polygons. A tetrahedron and a cube are the simplest examples of these. All of Euclid's many supporting diagrams can be drawn using only a compass and a straight edge.

The first printed edition was produced in 1482 and by some estimates *The Elements* is the most widely studied and translated publication in history after the Bible and the Koran. It remained the defining geometry text until the early twentieth century and earned Euclid the title, the Father of Geometry.

Little is known of Euclid himself. His very name means only "the famous man". The earliest biographies of him were written many centuries after his death and are most probably fictitious. Greek mathematicians including Archimedes, born only a few years after Euclid's death, acknowledged their debt to him. It became essential reading not only for geometricians but for any educated person. Abraham Lincoln carried a copy with him while studying to be a lawyer, because of the unshakable logic of Euclid's mathematical proofs.

*RIGHT: It is estimated that over a thousand different versions of Euclid's book have been printed since 1482. For centuries it was required reading for all university students. Fellow scientists Nicolaus Copernicus, Johannes Kepler, Galileo Galilei and Sir Isaac Newton were all influenced by what is a compendium of the work of the greatest Greek mathematicians.*

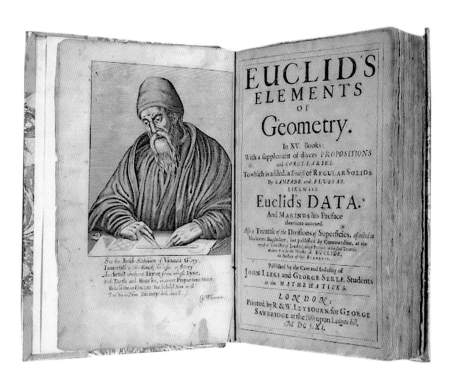

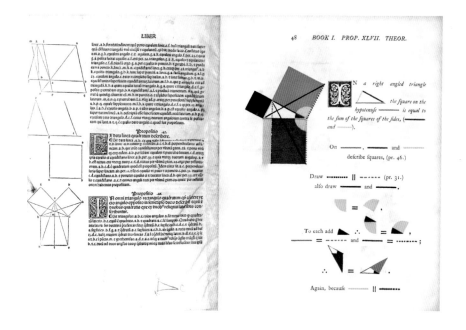

*TOP AND ABOVE LEFT: A textbook illustration of the Archimedes Screw used for irrigation, and the Romney Weir Hydro Scheme on the River Thames at Windsor, which uses Archimedes Screws to generate electricity.*
*ABOVE RIGHT: A mosaic of the death of Archimedes, killed by a Roman soldier during the siege of Syracuse.*
*OPPOSITE: An eighteenth-century portrait of Archimedes by Giuseppe Nogari.*

# Archimedes

## (c.287–212 BCE)

## The value of pi

Pi is the ratio between the width of a circle (its diameter) and its circumference, and between the width and the area. It is a constant – regardless of the length of the diameter and the circumference, Pi will *always* be the same number. But for centuries it remained to be calculated ...

In the precise, accurate world of mathematics, it is a source of some annoyance that pi, which is always the same, can never be accurate. It is, in arithmetical terms, irrational: it cannot be defined by a simple fraction such as 22 divided by 7. No matter how many decimal digits you calculate it to, there will always be more. Nor does it settle into a pattern of repeating numbers like, for example 10 divided by 3: 3.33333 33333 recurring. The first fifty decimal places of pi are 3.14159 26535 89793 23846 26433 83279 50288 41971 69399 37510.

The Babylonians knew about the constant four thousand years ago. A clay tablet from the period shows that they estimated it to be 25/8, or 3.16, which is pretty close. An Egyptian papyrus document of the same age puts it at $(16/9)^2$ or 3.125, closer still. By the fourth century BCE a Hindu text had refined it to 339/108, very close – 3.139 – to the approximate value used in classrooms today of 3.14.

In around 250 BCE the great Greek mathematician Archimedes came up with a way of refining the value of pi even further. The circumference of a polygon, with its straight lines, is easier to calculate than that of a circle. Archimedes placed one hexagon inside a circle and another outside it, and compared their circumferences; the circle's circumference must lie between those two measurements. The more sides a polygon has the closer it approximates to a circle; and by the time he made the calculations using 96-sided polygons, he could prove that pi lay between 223/71 and 22/7, a margin of error of only 0.0021, approximately.

Archimedes was the greatest mathematical mind of his age. Beside pi he also explored calculus and developed ways of calculating the surface area of a sphere and of less regular geometric shapes, such as a parabola and an ellipse. He put his theories to practical use, devising complex pulley systems and the famous screw pump, a refinement of a device used to draw water from rivers for irrigation. Today, the Archimedes Screw is still used in settings as varied as sewage farms and chocolate fountains.

Archimedes' value for pi persisted until 1630, when an Austrian geometrician, Christoph Grienberger, used the same polygonal technique to calculate pi to thirty-eight decimal places. By the end of that century English mathematician Abraham Sharp had applied a technique called Infinite Series to extend pi's definition to seventy-one decimal places. Within seven years John Machin, another Englishman, had broken the 100-place barrier.

Spare a thought for poor William Shanks, an amateur English mathematician who spent fifteen years in the mid nineteenth century calculating pi to 707 decimal places. After his death it was discovered that he had got the 528th number (and therefore all those that followed it) wrong. Nevertheless, his reduced record of 527 decimal places stood until the invention of computers a hundred years later.

# Eratosthenes

## (c.276–194 BCE)

## The size and shape of the Earth

It was the ancient Greeks, using their powers of observation, logic and deduction, who conceived that the Earth was a sphere (or to be more precise an oblate spheroid). And one man calculated its circumference with remarkable accuracy.

The great mathematician Pythagoras was the first man to theorize that Earth must be a sphere. In the sixth century BCE he puzzled over its cyclical relationship with the Sun and Moon; and a globe was the best explanation he could come up with. It took another three centuries for his fellow Greek thinkers to catch up, but by the third century BCE it was widely accepted.

A man known as Eratosthenes, from the Greek colony of Cyrene (now part of modern-day Libya), was appointed head librarian to the largest library in the world, in the Greek city of Alexandria, in 240 BCE. Eratosthenes was able to draw on Pythagoras's globe theory, the observations of others elsewhere in the world and Euclid's newly defined science of geometry, to answer a big question: how big is this sphere on which we live?

Eratosthenes arrived at the total by comparing the angles of elevation of the Sun at noon from different locations which were known distances apart – distances measured annually in Egypt for tax purposes. From these triangulations he calculated an answer of 252,000 stadia. One stadion, a standard Greek measurement of length, was based on the circumference of a stadium. Depending on which ancient stadium you measure, Eratosthenes' answer is accurate to within 2% of the actual measurement, 40,000 km (24,850 miles).

Armed with this knowledge he set about defining the known world and published his results in a trilogy of books. Although he wasn't the world's first mapmaker he is credited through this work with being the Father of Geography. He drew maps from which it was now possible to measure the distance between any of the four hundred cities that he described and plotted. He overlaid grids on his maps and devised early versions of latitude and longitude. He divided the world into five regions by climate: the two poles, two temperate regions and the tropics around the equator. It was the first time that the whole of the known world had been presented together in one rational and accurate format.

Not content with that achievement, Eratosthenes also measured with considerable accuracy the distances from the Earth to the Sun and the Moon. He fixed the length of the year at 365 days and created the four-yearly leap day as a corrective measure. Still on the subject of time, he attempted to write an accurate chronology of history – for example in his time the sacking of Troy was no more than an old legend shrouded in myth and embellishment. He came up with a date of 1183 BCE for it, which is within modern archaeological estimates of 1260–1180 BCE.

Eratosthenes was a prolific author, but none of his works have survived the intellectual purges and several fires that eroded the collection of the library in Alexandria. We rely on fragments quoted by other authors, and their tributes to him, to know his achievements.

*ABOVE RIGHT: Greek mathematician Eratosthenes' Well on Elephantine Island on the River Nile, from which he was able to calculate precise angles of the sun at various points of the calendar.*
*RIGHT: A nineteenth-century reconstruction of Eratosthenes' map of the world from 194 BCE.*

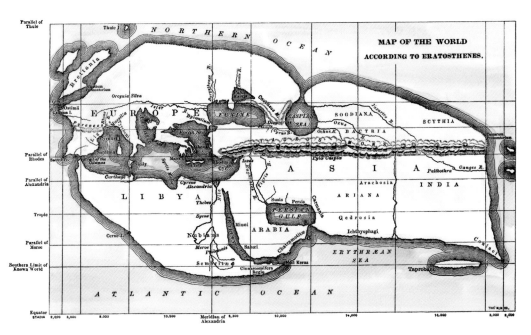

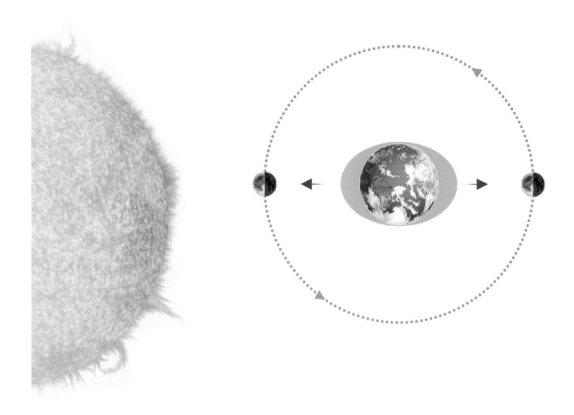

ABOVE: *An illustration showing that when the Moon and Sun are aligned (full and new moon) the biggest differences between high and low tides occur.*
OPPOSITE: *Luke Jerram's "Museum of the Moon" pulls in an audience at Dorchester Corn Exchange, on the 50th anniversary of the first moon landing, in 2019.*
LEFT: *The ancient ruins of Seleucia.*

# Seleucus, Seleucia

## (c.190–c.150 BCE)

## The Moon's effect on the tides

Seleucus of Seleucia patiently observed the changes in tides for years before coming to the conclusion that they were regulated by the Moon. He was right about the cause, even though he was wrong about the mechanism that created them.

An understanding of the tides is essential for all those living in coastal areas. Sailors have to know when it's safe to sail in shallow areas, and when the tidal currents are running strongest; landlubbers need to anticipate exceptionally high tides, which, combined with stormy weather, might inundate crops and homes.

It would be a simple if time-consuming process to watch and record the tides over a year – the phenomenon of two high and low tides a day is easy to establish; and even without a calendar it becomes clear that the full moon and new moon coincide with tides of greater range than at other times – spring tides. Seleucus thought it was more than mere coincidence and looked for a scientific explanation.

His own coast was either the Persian Gulf southeast of modern-day Iran, or the Erythraean Sea on the southeastern edge of Arabia – there are competing locations for his home town Seleucia. He complemented his personal observations with tidal records from other regions. He noted that tides varied in different parts of the world, and that tides were affected by the position of the Moon in relation to the Sun. And he attributed the very existence of tides to the attraction of the Moon.

Seleucus believed, correctly, that the Earth spun on its own axis and revolved around the Sun. He was wrong in one detail. He explained variations in terms of winds generated between the Moon and the revolving Earth. Although there are no such winds, they could, with hindsight, be seen as a metaphor for gravity, and for the changing position of the Earth in relation to the Moon and Sun.

We now know that the gravitational effect of the Moon is twice that of the Sun; so when the Moon and Sun are aligned, at the full moon and the new moon, the pull on the tides increases by around 50%, producing those extra-high spring tides. There are around 150 other factors which affect the range of tides including the shapes of the seabed and the coastline. The daily cycle of two tides takes around twenty-four hours and fifty minutes, so the time of each high tide is about fifty minutes later than the previous day's.

Heliocentrism, Seleucus's belief that the Earth moved round the Sun, was essential to his understanding of tides. He would not have been able to imagine his solution if he had believed the Earth to be the centre of the universe. The philosopher Plutarch (c.46–c.119 CE) said that Seleucus was the first to prove heliocentrism by logic, thanks to his observations of the planets.

It is thought that Seleucus drew on the relatively recent techniques of geometry and trigonometry to reach his model of the Solar System. Thus, like Sir Isaac Newton and all scientists, Seleucus might well have said, "If I have seen further it is by standing on the shoulders of giants." New science builds on the old.

# Ptolemy

## (c.100–c.168 CE)

## Predictions of planetary movements

Claudius Ptolemy's mathematical analysis of the stars and planets produced a model of the universe which endured for 1200 years. It was the product of a quarter-century of study and is the oldest comprehensive work of astronomy to survive.

Ptolemy lived in Alexandria after it had become part of the Roman Empire. Although his forename is Roman, his surname suggests that his family may have settled in the area when it was still under Greek rule. He wrote in Greek, not Latin; and although little is known about his life, a very great deal is known about his life's work.

Ptolemy was a prolific author on a wide range of sciences including geography, optics and harmonics – the mathematics of music. His works survive in copies or translations sometimes made hundreds of years after his death, proof of the depth of his knowledge and influence on the scholars who followed him.

His greatest contributions were in the field of astronomy, which he published as *The Mathematical Treatise*. Over time it became known as *The Great Treatise* and in Arabic, as *The Greatest*, "al-majisti" – hence the name by which it is known today, *The Almagest*.

*The Almagest* consists, according to a Latin translation of 1515, of thirteen sections or "books" divided by subject matter – such as the Sun, the Moon, the Stars, constellations, eclipses, the lengths of the day and year. The first book lays out Ptolemy's overview of the universe. He was a geocentrist: that is, he believed the Earth to be at the centre of everything, a sphere enclosed by the sphere of the heavens. This was not true,

but it was a natural assumption to make. Despite occasional claims to the contrary, it wasn't until Nicolaus Copernicus argued the case for the Sun as the centre that geocentrism finally gave way to heliocentrism.

Geocentrism did not affect the thoroughness of Ptolemy's observations. *The Almagest* was the result of at least twenty-five years of stargazing, all with the naked eye; and it was the definitive work on the subject until the invention of the telescope. It included a catalogue of the stars, 1022 of them, "as many stars," he wrote, "as it was possible to perceive"; and a record of forty-eight constellations that were visible from his region of the world.

Ptolemy plotted the positions of the stars and planets using the science of trigonometry developed by the Greek mathematician Hipparchus. Hipparchus was also an astronomer, and is credited with discovering precession – the fact that the angle of the axis around which the Earth rotates is changing. Ptolemy took this into account when publishing his own observations, and included an almanac whereby the future positions of the stars could be predicted. This handy guide was the standard reference work for European astronomers for the next fourteen hundred years. Ptolemy's comprehensive work was the benchmark against which all future cosmological discoveries were compared.

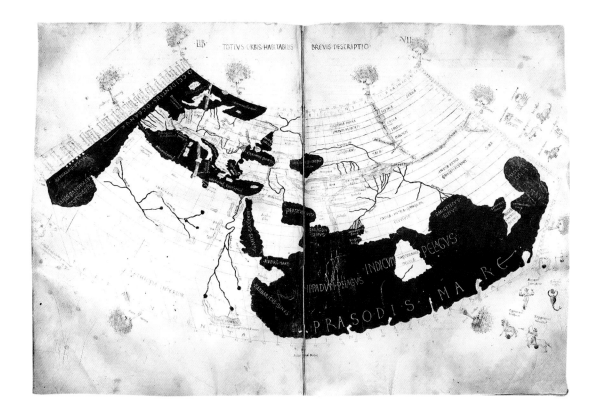

TOTIVS ORBIS HABITABILIS BREVIS DESCRIPTIO

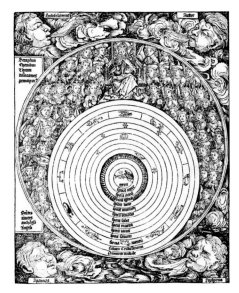

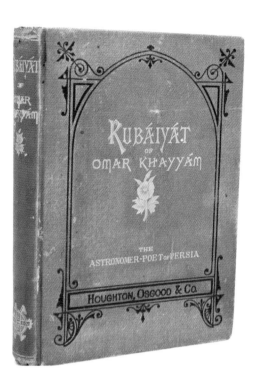

TOP: The Rubaiyat of Omar Khayyam *was first translated by Edward FitzGerald in 1859 and was a favourite of the Pre-Raphaelites.*

ABOVE: *Afghans gather to celebrate Afghan New Year (Nawroz) in Kabul, March 21, 2013. Afghanistan uses the Persian calendar, which runs from the vernal equinox.*

OPPOSITE: *There are many statues to the great Astronomer-Poet, this one in the city of Shiraz, Iran.*

# Omar Khayyam

## (1048–1131)

## The length of the calendar year

Omar Khayyam is best known today for his poetry, translated by Edward FitzGerald as *The Rubaiyat of Omar Khayyam* in 1859. Science historians remember him more for his astonishingly accurate calculation of the length of the calendar year.

Omar Khayyam's surname means "tent maker". We don't know if he was a maker of tents, but if he was, it would explain Khayyam's great ability with geometry and algebra: designing and fitting the different panels of even the simplest tent requires a grasp of shape, angle and size.

Khayyam was famous in his lifetime for his mathematical writings. He sought to unify the two apparently different disciplines of algebra and geometry and was one of the first thinkers to expand on Euclid's geometry treatise *Elements*. "No attention should be paid," he wrote, "to the fact that algebra and geometry are different in appearance. Algebras are geometric facts which are proved by propositions five and six of Book Two of *Elements*." Khayyam proved Euclid's Parallel axiom, and was the first to explore cubic equations, which he solved using geometry.

He lived in the north-eastern area of modern-day Iran, and his fame brought him to the attention of the sultan Malik-Shah, who engaged him in 1074 to build an observatory in the city Isfahan and recalibrate the Persian calendar.

The Persian calendar is one of the oldest records of time in history. Its earliest incarnations are more than 3,000 years old; the oldest surviving form dates from the fifth century BCE; and it has been subject to regular revisions. Broadly speaking it was based on phases of the moon, and changes in it were the result of political or climatic changes, or of the failure to use leap years: by 1079 the calendar, last revised in 895 CE, was out by eighteen days.

Khayyam worked with a team of eight astronomers making detailed observations of the cycles of the heavens and in 1079 he introduced the Jalali ("royal") Calendar. He began by making the start of the year and the start of the calendar the same day – the Spring Equinox, usually 21 March. In the past, the two events had both been variable. He abandoned the traditional monthly divisions directed by Zoroastrian festival days; instead the months changed with each zodiacal sign, and were of more equal length as a result.

Omar Khayyam's greatest contribution to the calendar was his calculation of the length of the year. The Gregorian calendar used in most countries, with 365 days for three years then 366 in a leap year, is pretty accurate – it will be a day out every 3,330 years. But Khayyam devised a cycle of thirty-three years, which is accurate to within a day every 5,000 years. The cycle includes twenty-five years, 365 days long, and eight leap years falling every four or five years.

The Jalali calendar remained in use as Khayyam devised it until the early twentieth century. It was simplified in 1925, and as the Solar Hijri Calendar it remains the official calendar of Iran and Afghanistan today.

Khayyam's poetry also showed a preoccupation with Time, or the lack of it. One of his *Rubaiyat* ("quatrains") reads:

Come, fill the cup and in the fire of Spring
The Winter garment of Repentance fling:
The bird of Time has but a little way
To fly – and Lo! the bird is on the wing.

# Shen Kuo

## (1031–1095)

## True north

In Europe he would have been called a Renaissance Man, a scientist who was also a poet, musician, civil servant, diplomatic negotiator, military tactician, cartographer and author. Shen Kuo was a polymath and a meticulous recorder of the scientific achievements of his countrymen under China's Song Dynasty.

Thanks to the writings of Shen Kuo we can appreciate just how technically advanced China was in the eleventh century. He noted, for example, the invention of movable type by Bi Sheng around 1040, some four centuries before Johannes Gutenberg introduced it to Europe.

In retirement Shen spent his twilight years on his country estate, called Dream Brook, near present-day Zhenjiang on the Yangtze river. There he spent time with what were known as the Nine Guests: meditation, conversation, painting, playing the zither, poetry recitation, alchemy, drinking tea, drinking wine, and playing the board game weiqi. He also found time to write his memoirs, the Dream Pool Essays, which have survived.

The Essays included chapters on anatomy, pharmacology, optics, civil engineering, archaeology, geology, climate change, meteorology, astronomy and even ufology – one passage describes regular visits by an unidentified flying object from whose open door a bright light shone, "like the rising Sun, lighting up the distant sky and woods in red". It was a local tourist attraction and a pavilion was built from which to see it.

They also contain the first known reference to a magnetic compass. The compass is one of the Four Great Inventions celebrated as evidence of China's superiority in technical development (alongside gunpowder, paper and printing). In Shen's lifetime, craftsmen had discovered how to magnetize iron needles by rubbing them with a piece of lodestone (magnetite). Before that, travellers had used iron weakly magnetized by being heated, and before that the south-pointing chariot, a third-century invention – a wooden pointer, mounted on a cart and aimed southwards before the start of a journey, maintained its direction by being geared to the cart's wheels.

Shen went into some detail about the best way to use a magnetized needle. He had made extensive observations with one and had made his own discovery about it. He was the first man to notice that a compass did not point to the south as indicated by the midday sun, but consistently slightly east of south. He had effectively discovered the difference between true and magnetic north and south.

He was then able to calculate the rate of declination and relate it to the stars by which sailors had always navigated. He was also the first to observe that the Pole Star, traditionally thought to be a fixed point directly over the North Pole, in fact revolved very slightly around it. These discoveries made accurate navigation possible and Shen Kuo was able to recommend the use of a compass with twenty-four points instead of the more common eight. The first recorded use of such a compass was shortly after Shen's death. The first use of a compass in the western world was about a hundred years later.

Shen's discovery of the variation between magnetic and true north may seem trivial, but the ability to travel accurately across great distances expanded the possibilities of international trade and opened up the world to cultural and economic exchange. The compass was the first basic steps towards the Global Village.

ABOVE: *A bronze bust of Shen Kuo on display in the garden of the Beijing Ancient Observatory. Apart from his work with the compass he improved the design of astronomical instruments and formulated a solar calendar named the Twelve Solar Terms Calendar, which was used to improve Chinese agriculture.*
LEFT: *An ancient Chinese compass.*

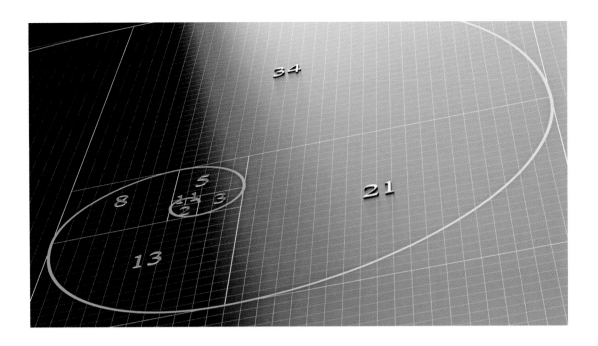

TOP: *A Fibonacci spiral using the Golden Ratio.*
ABOVE: *The cross-section of a nautilus shell shows a perfect Fibonacci curve.*

# Fibonacci

## (c.1170–c.1250)

## Numbers

Ancient Greece produced some brilliant mathematicians, the Roman Empire none at all. Neither culture had separate symbols for numbers, just letters from their alphabets, which made calculations clumsy and difficult. We must thank Leonardo Bonacci of Pisa for sorting out the mess.

We are still familiar with Roman numerals, which nowadays are used mostly for counting kings and queens – Louis XIV, Edward VIII, and so on. Rome used a handful of letters to denote key amounts; and anything in between was represented by multiple letters. So 2877 would be represented as MMDCCCLXXVII: two thousands (M), a five hundred (D), three one-hundreds (C), a fifty (L), two tens (X), a five (V) and two ones (I), all adding up to 2877.

The Greeks assigned the first nine letters of their alphabet to the numbers 1 to 9; the next nine to 10 to 90, the next nine to 100 to 900, and so on. When they ran out of letters they went back to the beginning, adding an extra mark to show that – for example – A (alpha) meant 1 but A' (with an apostrophe) meant 1000. This was scarcely more manageable than the Roman system, and both empires used an abacus for simple calculations. Sums on paper were problematic because their numerals did not line up in neat columns. Subtracting 99 from 100 for example would look like this to the Romans:

$$C$$
$$- LXXXXVIIII$$
$$=I$$

and this to the Greeks:
$$P$$
$$- Q\Theta$$
$$=A$$

Leonardo Bonacci, known as Fibonacci (short for *filius* Bonacci, "son of Bonacci"), was born in Pisa, northern Italy. His father was a merchant in charge of the trading post at Bugia in Algeria. There, in northern Africa, the young Leonardo discovered the Arabic system of numerals. Originally developed in India some time after the first century CE, it was adopted by Arab traders around 900 CE. Instead of letters it used separate symbols for the digits 1 to 9 and separate columns for tens, hundreds, thousands and so on. It's the system we use today and it is far more efficient than its predecessors.

As the son of a merchant Fibonacci could see at first hand the advantages of the Arabic system for ease of use, and he became evangelical about it. He wrote *Liber Abaci*, "the Book of Calculation", in 1202. It introduced the system, compared it to the traditional Roman one and demonstrated its use in bookkeeping, including calculations of profit and interest, and exchange rates. The book's popularity hastened the adoption of Arabic numerals by European commerce and assisted the development of banking, then in its infancy.

Although he didn't invent it, Fibonacci is best remembered today for the Fibonacci Sequence, featured in *Liber Abaci*, in which each number is the sum of the two previous ones: 0, 1, 1, 2, 3, 5, 8, 13, 21, 34, 55.

The numbers in the sequence are known as Fibonacci Numbers, and they occur in several mathematical and natural situations. Squares with sides of Fibonacci Numbers, arranged in descending order, produce the Fibonacci spiral, found in the uncurling of a fern and the structure of a pine cone, among other things. *The Fibonacci Quarterly* is an academic journal first published in 1963, still going strong and dedicated to examples of the occurrence of Fibonacci Numbers in nature and science.

# Roger Bacon
## (c.1219–c.1292)
# Optical theory

Roger Bacon conducted early experiments in the properties of light and eyesight. Building on studies from the Islamic world he laid down the foundations of the modern science of optics.

Hailed as the first modern scientist, Roger Bacon was a combative cleric who believed in empirical evidence rather than faith in the claims of others. He was determined to experiment for himself and believed that the wonders of science were the best proof of the existence of God.

Or so he argued, when he approached Pope Clement IV to ask for funding for an all-encompassing encyclopedia in 1266. Clement, unfortunately, misunderstood and demanded to see a work which he thought Bacon had already completed. In order not to disappoint His Holiness, Bacon hurriedly began to write; and within a year had written his *Opus Majus*, his great work – a summary of all his scientific knowledge including mathematics, alchemy, astronomy and optics.

It was an impressive achievement, written with astonishing speed and demonstrating a remarkable breadth of knowledge. Furthermore, it was written in secret. The Pope commanded Bacon not to reveal its contents to anyone until he, Clement IV, had seen them. And as a Franciscan monk, Bacon was expressly forbidden from writing books without permission. He had to write *Opus Majus* illicitly in his spare time.

Bacon came from a wealthy family in Somerset, England. Before he became a friar he had attended university where his subject was Aristotle. While Aristotle's teacher Plato believed that everyone and everything on Earth was an imperfect version of a heavenly ideal, Aristotle was much more down to earth, concerned about creating perfection in the here and now through philosophy. This concern with the fabric of the world we live in captured Roger Bacon's imagination and he began to collect books and read avidly on the natural sciences which he encountered when he returned to Oxford to teach in 1247 – the same subjects which later appeared in *Opus Majus*.

The first section of *Opus Majus* was titled "The Four General Causes of Human Ignorance". Bacon wanted to see things for himself, not simply accept the dubious claims of others. He was fascinated by lenses and mirrors. His use of primitive microscopes and telescopes, which he claimed "brought things near", earned him a reputation as a wizard. He terrified his students by a demonstration of refraction, in which he conjured up a rainbow by passing white light through clear glass beads. Before the invention of spectacles he suggested that imperfect human vision could be corrected with the use of lenses. And he once observed an eclipse of the Sun through a camera obscura.

Bacon's desire to conduct practical experiments instead of explaining the world through philosophy and religion made him a pioneer of the modern scientific method. In addition to his major work in optics, Bacon attempted to recreate the early chemistry of the alchemists; and he was the first westerner to describe how to make gunpowder, a century before the invention of the gun. He predicted ships that did not require wind, flight by balloons and flapping machines, and horseless carriages. He was a man about 300 years ahead of his time.

*ABOVE RIGHT: Optic studies from Roger Bacon's De multiplicatione specierum. The diagram shows light being refracted by a spherical glass container full of water.*
*RIGHT: A statue of Roger Bacon carrying an armillary sphere (as was Ptolemy) in the Oxford University Museum of Natural Science.*

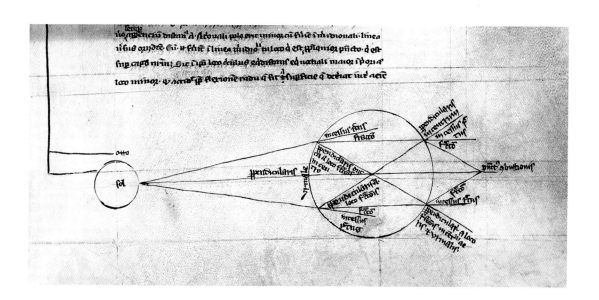

TOP: *The chapel of the Sorbonne University in Paris where Jean Buridan taught in the Faculty of Arts.*

ABOVE: *Aristotle believed there were two kinds of motion: "violent or unnatural motion", such as that of a thrown stone, and "natural motion", such as that of a falling object. Buridan argued that a force was transferred to the stone.*

# Jean Buridan

## (c.1301–c.1359)

## Theory of impetus

What keeps a stone in the air after you have thrown it? Aristotle believed that it was currents of energy in the air around the stone. French philosopher Jean Buridan suspected that a different force was at work, and he named it "impetus".

The details of Buridan's life and death are few, but he seems to have been a colourful character, based on the (completely unsubstantiated) stories that were told about him. One says that he died after being tied in a sack and thrown into the River Seine on the orders of the French king, having had a love affair with the French queen. Another claims that he and Pope Clement VI vied for the affections of a shoemaker's wife and that Buridan struck the pope a blow to the head with a shoe. Yet another records that, to escape philosophical persecution, Buridan fled to Vienna and founded the city's university.

Whatever the truth of such gossip, he was clearly a man about whom people told colourful tales. He built his reputation over a lifetime spent entirely at the University of Paris where – unusually for a philosopher – he taught exclusively in the Faculty of Arts instead of progressing, as was expected, to the Faculty of Theology. In doing so he drew a distinction between science and religion that was contrary to the prevailing doctrine of the time.

Buridan was a nominalist and had studied under the leading medieval nominalist William of Ockham. "Ockham's Razor" is the philosophical argument that the simplest explanation is most likely to be the correct one, and that one should not clutter matters with unnecessary assumptions. Perhaps this is why he questioned

Aristotle's theory of vibrating eddies as the invisible means of support for flying objects.

Instead, Buridan argued, a stone remained in flight because a force was transferred to it by the hand that threw it. He was not the first to consider an alternative to Aristotle's theory; Middle Eastern philosophers had considered something similar in the sixth and twelfth centuries. But Buridan was the first to coin the term "impetus" for the force imparted to the stone. And he was the first to claim that weight, wind resistance and gravity were responsible for the loss of impetus and the fall of the stone to the ground.

Buridan's impetus theory was the first step towards the modern understanding of inertia. His ideas were further developed by Leonardo da Vinci, Galileo and Isaac Newton. In his own lifetime, however, Buridan's nominalism was his undoing. It offended the established church, which embraced precisely the sort of unprovable abstract ideas that nominalism cast aside. An edict against nominalism by the French king Louis XI resulted in the banning of Buridan's books and may have been behind those gossipy rumours of a rift with the king and a flight to Vienna. According to Ockham's Razor one should not make unnecessary assumptions; but Buridan's death somewhere between 1358 and 1361 remains unexplained. Did he get the sack after all?

# Nicolaus Copernicus

## (1473–1543)

## A Heliocentric Universe

Although ancient Greek thinkers believed that the Earth orbited the Sun, Western Europe still believed in the Middle Ages that everything in the Heavens revolved around the Earth. Polymath Nicolaus Copernicus of Poland, however, found many questions unanswered by that assumption.

The big problem with the Earth being at the centre of things was that although most of the time the planets seemed to orbit in one direction, every now and then they would appear to start going backwards – to be retrograde, as the astrologers of the time described them. This seemed to Copernicus to be unlikely. There were other issues. It could not, for example, explain the variations in brightness of planets which were supposedly a constant distance from the Earth as they spun around it.

Copernicus was born in northern Poland and studied astrology in Krakow and Bologna. At the latter he lodged with the city's official astrologer, whose job was to interpret the influence of the planets on the health and fate of individuals. In Padua he studied medicine, a science then related to astrology because of the supposed effect of the planets on the human constitution.

His uncle, Bishop Watzenrode, found Copernicus a position as a canon, which gave Nicolaus plenty of spare time to pursue his astrological interests. His reputation was sufficient by 1515 for the Pope to invite him to contribute to the reform of the Roman Catholic calendar. The Church was still using the Julian Calendar, so called because it had been devised during the time of Julius Caesar 1600 years earlier; and it was by now considerably out of alignment with festivals based on the position of the Sun.

By then Copernicus had already proposed his ideas of a universe centred on the Sun – heliocentric, from the Greek word Helios ("sun") – in a short pamphlet known as the *Commentariolus*. It offered no proof, and it was a challenging and potentially heretical concept. Heaven was where the stars were. If Earth, with all its pain and suffering and cruelty, was not separate from Heaven but a part of it, then was Heaven itself tainted and imperfect?

Copernicus refined his theory, expanding it over time into a considerable body of text with dense mathematical proofs, and sharing it with friends. Word spread throughout European academia and at last, in 1542, he was persuaded to publish his book, *De Revolutionibus Orbium Coelestium* ("On the Revolutions of the Heavenly Spheres"). It is said that he was presented with a copy on his deathbed in 1543.

It was not universally well received. Even before its publication, theologian Martin Luther condemned it: "This fool wishes to reverse the entire science of astronomy; but sacred Scripture tells us that Joshua commanded the sun to stand still, and not the earth." Astronomers admired it, but the first edition run of 400 copies did not sell out. In the ensuing religious debate it was even placed on the Vatican's Index of Forbidden Books.

Nevertheless his theory caught on, pursued by Galileo Galilei and fully accepted by the time of Isaac Newton at the end of the seventeenth century. The Copernican Revolution in scientific ideas changed forever the view of mankind's place in the universe and created a permanent distinction between scientific knowledge and religious faith. Two hundred and seventy-six copies of the first edition of *De Revolutionibus* survive, compared to 202 copies of Shakespeare's First Folio and only 21 complete copies of the Gutenberg Bible.

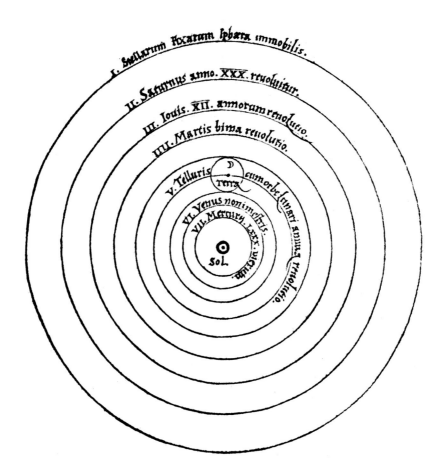

ABOVE: *Copernicus's revolutionary view of the universe was crystallized in this simple yet disconcerting line drawing. His heliocentric model – which placed the Sun and not the Earth at the centre of the universe – contradicted fourteen centuries of belief.*

LEFT: *A hand-coloured engraving from Andreas Cellarius's* Harmonia Macrocosmica *(1708), showing the orbits of the known planets around the Sun with a representation of Copernicus at bottom right.*

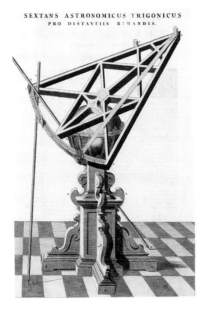

SEXTANS ASTRONOMICUS TRIGONICUS
PRO DISTANTIIS RIMANDIS.

LEFT: *One of the last astronomers to work without a telescope, Brahe insisted on the most accurate sextants and quadrants, which he had built in large form.*

BELOW: *Uraniborg as drawn by famous Dutch mapmaker Willem Blaeu in 1663.*

OPPOSITE: *Tycho Brahe was an imposing figure with a distinctive moustache. He lost part of his nose in a drunken sword fight and was forced to wear a prosthetic nose for the rest of his life.*

ORTHOGRAPHIA PRÆCIPVÆ DOMVS ARCIS VRANIBVRGI
in Infula Porthmi Danici Venufia, *Vulgo* Huenna, Aftronomiæ inftaurandæ gratia, circa annum MDLXXX.
à TYCHONE BRAHE ædificatæ.

# Tycho Brahe

## (1546–1601)

## Observation of a supernova

Until he saw it for himself, the fourteen-year-old Tycho Brahe could not believe that it was possible to predict a solar eclipse. The event, on 21 August 1560, prompted a lifelong study of the stars from which the first accurate maps of the heavens emerged.

Tycho Brahe was the son of a Danish nobleman, and in keeping with the practice of the age he began his university education at the age of twelve. He studied law, languages and the arts, first in Copenhagen and then in Leipzig.

Witnessing the solar eclipse fired Brahe's imagination: its predictability suggested that there was order in the movement of the heavens. He began to make and record his own observations and to collect books on the subject, including the predictive almanacs of the great early astronomers Ptolemy and Copernicus.

He observed a predicted alignment of Jupiter and Saturn in 1563, but found that both Ptolemy and Copernicus were out in their forecasts – Ptolemy by almost a month. Brahe, at the tender age of seventeen, determined to improve on their work. In pursuit of accuracy he set about acquiring the finest instruments available for his observations. If they weren't up to the task, he also designed his own. In pursuit of knowledge he toured the universities of Europe meeting fellow astronomers, sharing and gaining information.

From an ad hoc observatory in Denmark's Herrevad Abbey he observed the event which secured his scientific reputation. Looking up at the constellation of Cassiopeia in the autumn of 1572, he saw a luminous star which had not been there before, shining brighter than Venus. He observed it for a full year before publishing his findings as *De Stella Nova* ("About the New Star").

Tycho Brahe had witnessed a supernova – the explosion of a star

*Tycho-Brahe,*
*Mathematicæ Astronome, Chimiste et*
*Seigneur de Knutstrop Mort a Prague in*
*1601*
*age 55*
*a Paris chez Daumont rue St Martin*

before it disappears, or collapses into a neutron star or a black hole. He was lucky; to this day the last recorded observation of a supernova in action was in 1604, by Brahe's gifted student Johannes Kepler.

Brahe's conclusions impressed astronomers throughout Europe who now beat a path to his door. Frederick II, the king of Denmark and a personal friend, gave Brahe an island on which to construct Europe's first purpose-built observatory. It cost a full 1% of Denmark's national budget, and Brahe named it Uraniborg after Urania, the astronomers' muse.

In an age before the invention of the telescope, Brahe's observations were made with the naked eye. He built a giant sextant to plot the positions of more than 1,000 stars with ten times greater accuracy than anyone had done before, to within a minute of an arc; and he marked them in brass on a huge wooden globe. For even greater accuracy he constructed Stjerneborg, an underground observatory beside Uraniborg, to protect his instruments from wind and other meteorological interference.

Interference came in the end from Frederick II's successor Christian IV, who gradually withdrew funding from an operation in which he was not interested. Brahe moved to Prague, where he passed on his passion and knowledge of astronomy to Johannes Kepler and where, in 1601, he died, having set new standards for precision and accuracy. Kepler carried his work forwards, and a double statue of the two men stands in Prague today.

# Johannes Kepler

## (1571–1630)

## The elliptical orbit of planets

German astronomer Johannes Kepler was a devout Christian who believed that God ordered the universe with perfect mathematics. He dedicated his life to proving it with precise measurements of the heavens, and in the process confirmed the heliocentric theories of Nicolaus Copernicus.

Kepler studied to become a theologian, but decided that his facility with numbers was the best way in which he could serve his god. In Kepler's time there was no distinction between science and other areas of knowledge. What we call science was until the nineteenth century merely "natural philosophy", one of many branches of philosophy. Astronomy was called astrology and included divination and prediction. Everything was seen through the prisms of the prevailing theological and philosophical discourses of the day; and in sixteenth-century Germany that the Moon, the Sun, the planets and the stars revolved around the Earth.

Nicolaus Copernicus challenged that geocentric view with his 1543 work *De Revolutionibus Orbium Coelestium* ("On the Revolutions of the Heavenly Spheres"), although his printer – anxious to avoid controversy – added a disclaimer. Kepler's astrological tutor Michael Mästlin had to keep his admiration of Copernicus quiet, but when Mästlin showed him a copy of *De Revolutionibus*, Kepler knew instinctively that Copernican heliocentrism had the ring of truth about it.

Mästlin recommended Kepler to Tycho Brahe, the leading astronomer of the day, who was looking for an assistant. Brahe was a geocentrist and assigned Kepler the job of making sense of Mars – under geocentrism Mars appeared to go backwards periodically during its orbit. Kepler built on Brahe's own observations and after Brahe's death Kepler continued what he called his "War with Mars".

Kepler made thousands of calculations. By placing the Sun at the centre of things instead of the Earth, Kepler at last made sense of Mars' orbit as it appeared from the Earth's perspective; and he was able to predict its future position in the sky with such accuracy that it bears comparison with modern results.

Not only did he prove Copernicus's theory, but he established that the planet's orbit was elliptical, not simply circular. He also observed that, despite the variable distance of Mars from the Sun, an imaginary line drawn from the Sun to Mars crossed equal areas of space in equal periods of time as the planet went round in its orbit. Kepler found these two attributes to be universally true when he observed the other six planets known at the time, and they now form Kepler's First and Second Laws. Kepler is believed to have invented the word "orbit".

These laws of planetary motion changed the way in which space was understood and explored. Being able to predict the positions of planets, and confirm those predictions by observation, gave astronomers a solid basis from which to make other observations. Just as Kepler's interpretation would not have been possible without Copernicus, so Sir Isaac Newton's work depended on the progress made by Johannes Kepler. Carl Sagan called Kepler "the first astrophysicist and the last scientific astrologer".

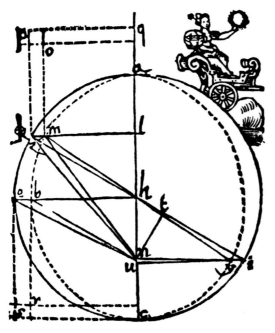

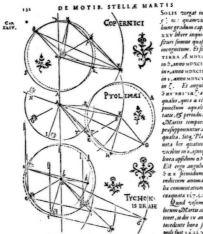

ABOVE: *This illustration from* Astronomia Nova *showed that the orbit of Mars is elliptical. Unlike the perfect circles of Copernicus's heliocentric model, Kepler deduced that the planets followed an elliptical orbit around the Sun.*

LEFT: *A page taken from Kepler's* Astronomia Nova *published in 1609.*

OPPOSITE: *A portrait of Kepler from 1610.*

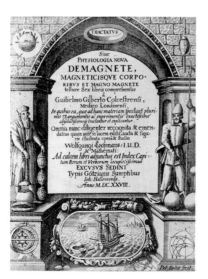

ABOVE: *A portrait of William Gilbert, who died in 1603, most likely of the bubonic plague.*
LEFT: *De Magnete, Magneticisque Corporibus, et Magno Magnete Tellure ("On the **Magnet** and Magnetic Bodies, and on the Great Magnet the Earth") published in 1600. Gilbert was the first to argue that Earth had an iron core.*
OPPOSITE: *Illustration of a declinometer from* De Magnete.

# William Gilbert

## (1544–1603)

## The Earth's magnetic field

William Gilbert was the leading English scientist, or "natural philosopher", of his day and served as Queen Elizabeth's personal physician. He pioneered experimental investigation, and his treatise on magnetism, published in 1600, is regarded as one of the first significant English works of science.

Gilbert studied medicine in Cambridge and practised as a doctor in London and throughout Europe for most of his life. After being elected president of the Royal College of Physicians in 1600 he was appointed to guard Queen Elizabeth's health in her final years. After her death he would have continued to serve her successor, James I of England and VI of Scotland, had he not himself succumbed to the bubonic plague in 1603.

In the course of his European travels he became interested in the magnetic properties of lodestone (the mineral magnetite), which predated the compass as a means of fixing direction. It is the most magnetic naturally occurring mineral, a complex oxide of iron known as ferrous-ferric oxide. The fact that it attracted iron objects was really all that was known about magnetism before Gilbert. His book, *De Magnete, Magneticisque Corporibus, et Magno Magnete Tellure* ("On the Magnet and Magnetic Bodies, and on the Great Magnet the Earth") was a summary of existing understanding of the subject, to which he added the results of his own observations.

Compasses had begun to appear in Europe during the previous century, although they had been in use in China for some four hundred years before that. Having noticed that a compass needle curved downwards slightly, he theorized that the Earth acted as a giant magnet on it. He was the first man to propose that the Earth had a magnetic field, with which the needle of a compass aligned itself north–south, and he invented the terms North and South Pole.

We still use the term "pole" in an electrical context, and Gilbert is sometimes considered to be the founder of electrical studies. Indeed, he coined the word "electricity" and was the first man to use terms like electric force and electric attraction. Gilbert drew a clear distinction between two forces of attraction; magnetism and static electricity (then known as the amber effect because amber acquired static electricity when it was rubbed).

These were early days for science, then still regarded as a branch of philosophical thought rather than a separate discipline; but Gilbert stressed the importance of practical experimentation and developed his own magnetic philosophy. Thus, when considering the movement of the tides, for example, he wrote in terms of "subterranean spirits and humours, rising in sympathy with the Moon, caus[ing] the sea also to rise and flow to the shores and up rivers".

Gilbert claimed that magnetism was the Earth's soul (where we might use the word "core"). He wrote of a lodestone carved into a perfect sphere, which would spin about its axis if correctly aligned with the poles. This analogy with the Earth was a controversial idea; it suggested that the Earth was not the fixed centre of the universe but rotated.

Gilbert's work was often overlooked in the two centuries following his death; but it attracted new interest in the nineteenth century when the science of electricity was rapidly developing. Although now obsolete, a unit of magnetomotive force was named the Gilbert (Gb) in recognition of his discovery of the Earth's magnetic field.

# Galileo Galilei

## (1564–1642)

## Laws of falling bodies

Galileo Galilei of Pisa was the consummate scientific all-rounder. Not for him the single-minded pursuit of knowledge in one specialized field: he studied widely across many disciplines including astronomy and engineering, hydrostatics and kinematics.

Galileo lived in an age when the natural world was still explained as the creative gift of God, and where the alchemical conversion of base metal into gold seemed perfectly possible. Anyone who suggested otherwise was open to accusations of heresy. Rational thought was a threat to the power of the church, which clung to the belief that God, and the Earth, were at the centre of all things.

Galileo is most commonly remembered today in the Galileo space telescope that bears his name. He made discoveries in several areas of science: he was, for example, the first man to observe Jupiter's four largest moons, and to realize that the Milky Way was not just a nebulous milky trail across the night sky but a densely packed band of stars. But what really marked Galileo out was his openness to the scientific study of motion and gravity.

The simple act of throwing a ball is governed by laws of science, which Galileo was the first to define. He challenged the Aristotelian orthodoxy, which believed that a heavier object will fall faster than a light one and proved that balls of different masses fell at the same speed.

The story that he demonstrated this by dropping objects from the top of the leaning Tower of Pisa is probably a myth – he makes no reference to it in his notes. He released balls at the top of an inclined plane to slow their speed uniformly and make it easier to measure accurately.

Aristotle thought that inertia was the natural state and believed that any moving object would eventually stop moving once the force that had propelled it stopped. Galileo recognized that what stops movement is not the absence of force but the presence of friction, without which moving objects retain their speed. Furthermore, without resistance, falling objects accelerate (he observed) at a uniform rate, which he correctly deduced is proportional to the square of the length of time of acceleration.

Galileo's studies of falling objects were rigorous and pioneering investigations which laid the groundwork for our understanding of gravitational force today. They are imperfect in that they do not take into account air resistance or the extremes of shape and size. For example, a feather weighs almost nothing but has a relatively large surface area with which to experience air resistance. From very great heights objects eventually reach a terminal velocity – that is, acceleration no longer increases at Galileo's uniform rate.

But in general his discoveries apply to all compact, dense objects falling from your hands, the Tower of Pisa or any other much taller structure which mankind might have built since. Astronaut David Scott demonstrated the truth of Galileo's science during the Apollo XV mission, when in the airless atmosphere of the Moon he dropped a feather and a hammer simultaneously and they both hit the moon surface together.

Galileo fell foul of the Spanish Inquisition, which in 1633 found him "vehemently suspect of heresy" for his support of heliocentrism. He was forced to "abjure, curse and detest" this truth, and was held under house arrest for the remainder of his life.

*RIGHT: A portrait of Galileo from 1636 holding one of his self-designed telescopes.*
*BELOW: Handwritten notes from the master on his astronomical observations.*
*BELOW RIGHT: The Apollo 15 mission where astronaut David Scott was finally able to prove Galileo's hammer and feather argument correct.*

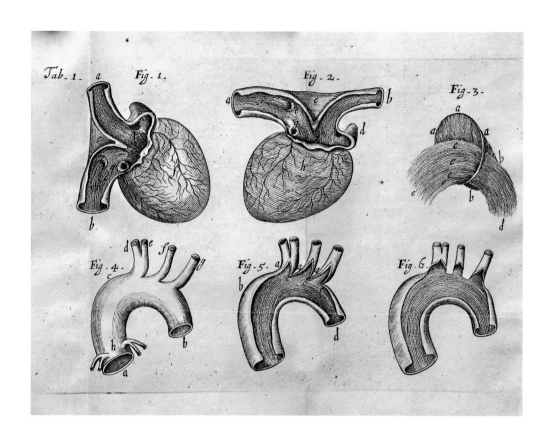

# William Harvey

## (1578–1657)

## Blood circulation

The heart has always been considered the physical centre of life, empirically demonstrated by its thumping beat in times of heightened stress. Its precise function, however, remained little understood until a modest English physician took a fresh look at human anatomy.

In the sixteenth century, "modern" medical practice was based on the research of Galen of Pergamon, a pioneer of medical science some 1500 years earlier. Although he made important discoveries in the fields of neurology, pharmacology and anatomy, his understanding of the heart was incomplete. Galen thought that the heart generated heat and was cooled by the lungs thanks to arteries which inhaled air and exhaled "vapours" through pores in the skin.

By the sixteenth century, physicians believed that there were two blood systems in the body. The arteries carried blood and "spirit" made in the heart, delivering heat and life to all corners of the body. Veins carried the cooler and darker venous blood, which was produced by the liver. The Galenic idea, that the lungs cooled the heart, persisted.

William Harvey was the eldest of nine children of the Mayor of Folkestone, a town on the south coast of England. After a general education in Canterbury and Cambridge he studied at Padua University in Italy, where he proved to be an exceptional student of anatomy under the great physician Fabricius.

Validated by degrees in medicine from both Padua and Cambridge, and a fellowship of the Royal College of Physicians, Harvey was invited to deliver the Lumley Lectures, a series of demonstrations intended to advance the knowledge of anatomy in England. His day job was as a physician in St Bartholomew's Hospital in London, an institution dedicated to healing the poor without reward.

EXERCITATIO ANATOMICA DE MOTU CORDIS ET SANGUINIS IN ANIMALIBUS. Cui accedunt EXERCITATIONES DUAE ANATOMICAE DE CIRCULATIONE SANGUINIS Ad JOANNEM RIOLANUM Filium; In Academia Parisiensi Anatomes & Herbariae Professorem Regium, Reginae Matris Lauberti XIII. Medicum Primarium. Auctore GULIELMO HARVEO Anglo, Anatomiae & Chirurgiae in Collegio Medic. Lond. Professore, Serenissimeque Maiestatis Regiae Archiatro. Hujusque Operum PARS PRIMA. EDITIO NOVISSIMA. Indice ornata. LUGDUNI BATAVORUM. Apud JOHANNEM van KERCKHEM. 1737.

Harvey conducted anatomical dissections of animals – fish, snails, birds, and the carcasses of deer killed on hunting trips – before confirming his theories on human corpses and finally on living humans. Without a microscope his only aid was a small magnifying glass. Some of his conclusions were necessarily inferred rather than definitively demonstrated; but they were all correct.

When it came to publishing his ideas, Harvey chose to do so in Frankfurt, which had hosted a book fair since the twelfth century and from which Harvey knew that his ideas, like blood, would be quickly circulated. His book *Exercitatio Anatomica de Motu Cordis et Sanguinis in Animalibus* ("Anatomical Account of the Motion of the Heart and Blood") contained the first accurate account of the circulation of blood in the body.

Among his then groundbreaking revelations were the facts that the arteries and veins were part of the same system, and that the left and right ventricles of the heart worked in tandem (not, as previously thought, independently). By simple mathematics he proved that it would be impossible for the liver to produce the amount of blood pumped out by the heart, nearly 500 pounds or 225 kilograms a day. Blood must therefore be circulated.

Nevertheless, *De Motu Cordis* met with considerable resistance, challenging as it did the Galenic principles of 1500 years. Despite the evidence and logic of Harvey's claims, many doctors said they "would rather err with Galen than proclaim the truth with Harvey". It took twenty years for Harvey's truth to be accepted.

# Pierre de Fermat

## (1607–1665)

## Fermat's Last Theorem

Was Fermat having a joke? When he first jotted down his "last theorem" in the margin of an old mathematical textbook in 1637, he added, "I have discovered a truly marvellous proof of this, which this margin is too narrow to contain." It took later mathematicians 357 years to find it.

Pure mathematics is to simple arithmetic as ballet is to the basic act of walking. In the minds of the great mathematicians, equations are the threads which prove the interconnectedness of all things, and French lawyer Pierre de Fermat was the greatest mathematician of his age.

The idea that the world can be explained in numbers is very old. The earliest example is perhaps Pythagoras's theorem that in a right-angled triangle "the square on the hypotenuse equals the sum of the squares on the other two sides". In mathematical terms that means that $a^2+b^2=c^2$, where a and b are the lengths of the two shorter sides of the triangle. When a, b and c are whole numbers they are known as Pythagorean triples. 3/4/5 is the best-known example – $3^2+4^2=9+16=25=5^2$ – and was used by early builders to construct perfectly right-angled corners.

There's a pleasing simplicity about $a^2+b^2=c^2$ which encouraged mathematicians to look for similar equations in which solutions could be found using only integers (whole numbers). Diophantus, who lived in Alexandria in the third century, was one such; Fermat was another. It was while reading Diophantus's great work *Arithmetica* that Fermat was inspired to write his own theorem in the margin – that an+bn=cn is impossible where n is any integer higher than 2.

His boast that he had a marvellous proof was not backed up with any evidence. If he did have any, it was not found after his death, when his son published the theorem for the first time. Fermat did prove it for the specific and limited case where n=4, in such a way that all anyone had to do now was prove that it was impossible for odd numbers.

That was easier said than done. Fermat's Last Theorem holds the title of the world's most difficult mathematical problem by dint of the sheer number of esteemed mathematicians who failed to prove it since Fermat scribbled it down. The proof, when it came, was further evidence that all things are connected by mathematics. First, in 1955, two Japanese mathematicians, Goro Shimura and Yutaka Taniyama, suggested a link between elliptic curves and modular forms, two previously unrelated fields of mathematics. Then, in Germany in 1984, Gerhard Frey saw a connection between modular forms and Fermat's equation.

The relationship between Fermat's theorem, modularity theorem and elliptic curves meant that proof of one would be proof of the others. It fell to English mathematician Andrew Wiles to provide the final proof. He had been fascinated by Fermat's Last Theorem as a child, and had also worked on elliptic curves. He worked in secret for six years, and his solution took up the whole of the May 1995 edition of the respected Princeton mathematical journal *Annals of Mathematic*.

In the best scientific tradition, Wiles' proof came to him in a eureka moment. His reaction to the discovery was one that could only have come from a pure mathematician. "It was so indescribably beautiful," he recalled. "It was so simple and so elegant. Nothing I ever do again will mean as much."

VARIA OPERA
**MATHEMATICA**
D· PETRI DE FERMAT,
SENATORIS TOLOSANI.

Accesserunt selectæ quædam ejusdem Epistolæ, vel ad ipsum à plerisque doctissimis viris Gallicè, Latinè, vel Italicè, de rebus ad Mathematicas disciplinas, aut Physicam pertinentibus scriptæ.

**TOLOSÆ,**
Apud JOANNEM PECH, Comitiorum Fuxensium Typographum, juxta Collegium PP. Societatis JESU.

M. DC. LXXIX.

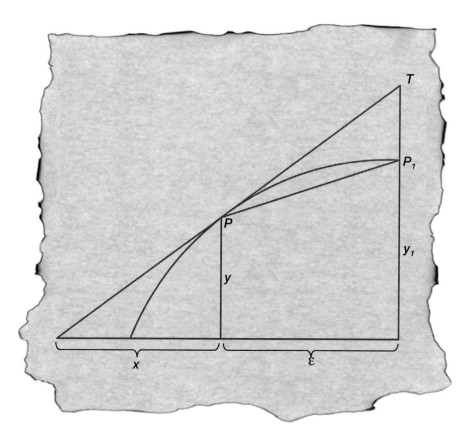

*ABOVE: Newton's Cradle, a device that demonstrates conservation of momentum and energy. Small steel balls work well because they remain efficiently elastic with little heat loss after strong strikes.*

*LEFT: A portrait of Newton from 1689.*

# Isaac Newton

## (1642–1727)

## Laws of motion

The apple tree still grows from which Isaac Newton's apple famously fell, at the scientist's home, Woolsthorpe Manor in Lincolnshire. Newton intuitively saw gravity at work and set about defining the forces acting on the fruit. In doing so he laid the foundations of classical mechanics.

Newton's three-volume masterwork *Philosophiæ Naturalis Principia Mathematica*, published in 1657, is regarded as one of the most important books in the history of science. *The Mathematical Principles of Natural Philosophy* (to give it its translated title) contained the first publication of his three Laws of Motion. They transformed our understanding of everyday movement and made Newtonian physics the dominant model until the emergence of Einstein's Theory of Relativity two hundred years later.

Put simply, the First Law states that if an object is at rest (not moving) it will remain at rest unless a force makes it move; and if an object is moving, it will continue to move unless a force stops it.

The Second Law says that the harder the force which pushes or pulls an object, the further and faster it will move.

The Third Law is perhaps the best known, that for every action (or force) there is an equal and opposite reaction. This is the law illustrated by that perennial desktop toy Newton's Cradle.

*Principia* also contained Newton's Law of Universal Gravitation. It describes the force of gravity which exists between any two particles in the universe, and mathematically defines it in terms of their masses and the distance between them. It became known as the First Great Unification because it tied together everything that was known about both gravity and astronomical movement. Such events in scientific history are not as common as you might expect – the Second Great Unification was James Clerk Maxwell's study of electromagnetism in the 1860s; the Third was Albert Einstein's early-twentieth-century unification of space

and time, and of mass and energy. The Fourth and most recent is Quantum Field Theory.

Newton's Laws of Motion are not just theoretical statements of the obvious. They are crucial in enabling us to predict the movement of objects under given conditions, whether it be the trajectory of a bullet or the power needed to lift a Saturn Five rocket into space. Newton used them to explain tides, and to confirm Johannes Kepler's observations of planetary motion, and Kepler's and Galileo's definitions of inertia.

Newton's Laws are now known not to be completely universal in their scope or accuracy. They do not apply to extremes of mass, speed or gravity, for example among particles smaller than an atom or inside a black hole. In such situations the Theory of Relativity holds sway; but our everyday lives and those of our earliest ancestors on Earth are still governed by the rules of movement that Newton first defined in 1687.

ABOVE: *Today, Woolsthorpe Manor with its famous apple tree is administered by the National Trust.*

# Robert Boyle

## (1627–1691)

## Law of gases

At a time when our world was still held to consist of only four elements – earth, air, fire and water – Robert Boyle, the son of an Anglo-Irish Lord, had his doubts. His experimental studies of gases, and Boyle's Law which resulted from them, earned him the title the Father of Modern Chemistry.

As a boy Robert Boyle received an education typical of the English landed gentry at the time. He was schooled at Eton and then embarked on a tour of classical Europe in the company of his French tutor. In 1641 they found themselves in Florence, Italy. There Boyle marvelled at the revolutionary scientific ideas of Galileo Galilei, who was frail but still alive and receiving visitors in his villa (he died the following year).

When Boyle returned to England in 1644 he dedicated the rest of his life to scientific research from his home in Dorset. He corresponded with others looking into this "new philosophy" as science was becoming known. They described themselves as an "invisible college", a virtual meeting of scholarly minds, which in 1663 was legitimized as the Royal Society, Britain's premier scientists' club. Boyle was a member of the Society's first council.

In the course of correspondence Boyle heard about an air pump created by a fellow experimenter, the German inventor Otto von Guericke, and set about improving its design. With the resulting device, which he modestly called the machina Boyleana (Boyle's Machine), Boyle began experiments to discover the properties of air. He published his initial findings in 1660 as *New Experiments Physico-Mechanical, Touching the Spring of the Air, and its Effects*.

The concept of air pressure was already known, but it was believed that air, and other gases, consisted of small particles surrounded by tiny invisible springs. When a Jesuit challenged Boyle's publication he responded with the equation that has become known as Boyle's Law. It states that, for a constant mass of gas at a constant temperature, the pressure is inversely proportional to the volume – in other words, the greater the pressure, the smaller the volume.

Although others earlier in the century had noticed something similar, it was Boyle's experimental proof that confirmed it. His Law was the first major one to be expressed as a mathematical equation. It is a cornerstone of classic mechanics in its application to pneumatic machines.

Boyle also made groundbreaking studies in genetics, concluding that all mankind shared a common ancestry and that physical characteristics were defined by the act of conception. He was unusual among his Anglo-Irish peers in believing that the Irish language was worth keeping and in the 1680s he funded the production of a complete edition of the Bible in Gaelic.

Not everything about Boyle was modern, however. Although he held enlightened views about human equality, he was a devout Christian and believed that the source of all mankind was Adam and Eve, whom he declared to be Caucasian. And despite his firm belief in evidence-based science instead of mere hypotheses, he was a committed alchemist who still believed that base metals could be transmuted into precious ones.

*RIGHT: An illustration of the apparatus Boyle used for experiments with air, taken from his publication* A Continuation of New Experiments Physico-Mechanical, Touching the Spring and Weight of the Air, and its Effects *(1669). This was a continuation of an earlier work of 1660, and was followed by a second part in 1682.*

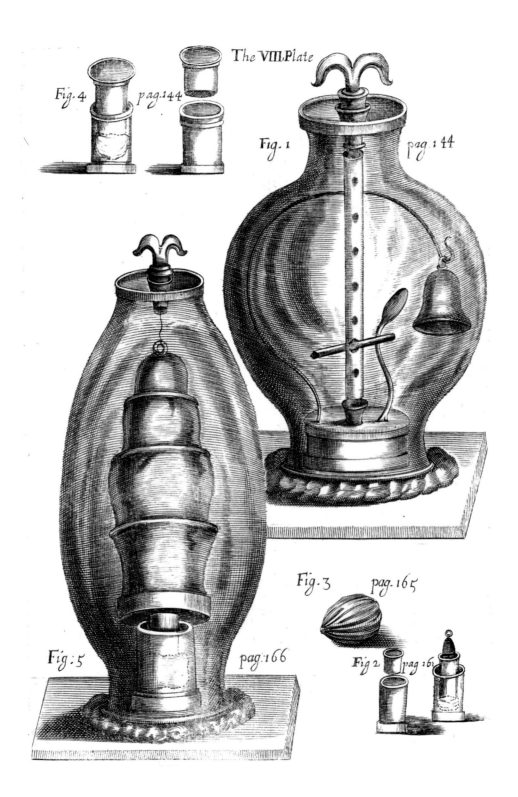

The VIII.Plate

Fig. 4        pag.144

Fig. 1        pag. 144

Fig. 3        pag. 165

Fig. 2        pag. 161

Fig. 5        pag. 166

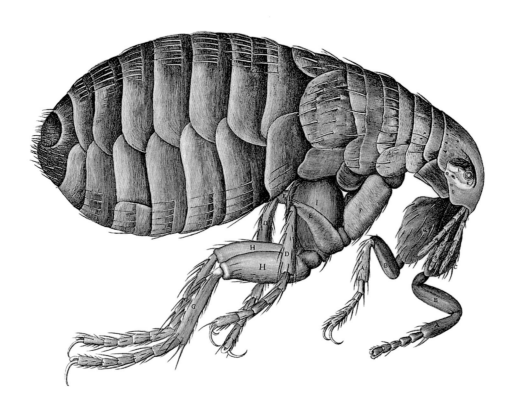

# Robert Hooke

## (1635–1703)

## The biological cell

Robert Hooke was part of the blossoming of science in England in the seventeenth century, a direct link between Robert Boyle and Isaac Newton and an innovator in his own right. His observations changed the study of several branches of science.

Robert Hooke's scientific career began with a job as assistant to Robert Boyle, for whom he built instruments and assisted with experiments that resulted in Boyle's Law of Gases. Boyle released him from that position so that he could take up a new one, created especially for him, as Curator of Experiments at the newly founded Royal Society. There he worked with the finest scientific minds of the age and presented experiments and ideas of his own.

It was during this time that he fell out with Isaac Newton, a future president of the Society, over who should have priority as discoverer of various phenomena including gravity. It has been suggested that Newton hid Hooke's papers and even got rid of the only known portrait of him. In recent times Hooke has been brought out from the shadow which Newton cast over him.

Hooke's insights were complimented by a rigorous mathematical and mechanical mind. As a child he is said to have dismantled a brass clock and built a working model of it in wood. In adult life he developed the use of the pendulum in clocks, and devised the balance spring, which greatly improved the accuracy of clocks and watches. It was he who suggested that an accurate timekeeper would solve the great maritime problem of establishing longitude.

The balance spring was the product of his experiments with springs and weights, which also produced Hooke's Law of Elasticity. The Law is an equation that connects the stiffness of a spring and the length to which it is stretched with the force needed to stretch it. It can also be applied to other elastic situations such as the bending of a guitar string or the movement in high wind of a tall building. The Law has applications throughout science and engineering and forms the basis of several fields of study including acoustics and seismology.

In a very different field of study, Robert Hooke published the book *Micrographia* in 1665. It contained the first ever drawing of a micro-organism, as observed through a microscope built to his design. His studies of microscopic animal and plant life led to his discovery of the cell, the smallest element of life. Hooke coined the term "cell" for these units because they reminded him of the cells in a honeycomb. The book contains an illustration of the cells in a piece of cork, like bricks in a wall, literally the building blocks of life.

Hooke was a polymath and has been described as England's Leonardo da Vinci. He also studied fossils at a time when extinction was a theologically unacceptable idea because it suggested fallibility. His works on the mechanics of human memory, published posthumously and overlooked for two centuries, are astonishingly modern in their understanding.

In his spare time Hooke was an architect and chief assistant to his friend Christopher Wren, with whom he undertook the rebuilding of London after the great fire of 1666. The dome of Wren's crowning glory St Paul's Cathedral was constructed according to Hooke's plans.

# Nicolas Steno

## (1638–1686)

## Fossil theory

Most people "knew" in the early seventeenth century that fossils were things that grew in the rock in which they were found. Others thought that they were embedded there having fallen from the Moon. Danish physician Nicolas Steno decided to dig a little deeper.

Curious minds in ancient Greece and medieval China saw fossils of recognizable organisms such as shells and plants as evidence of changing climates and shorelines. Less identifiable fossils seemed to bear no relation to living things and therefore could not be of this world. In Europe they were consigned to the curio cabinet and aroused little genuine curiosity about their nature.

Leonardo da Vinci was, as in so many fields of human endeavour, the exception. His studies of the fossils of animals and their burrows or bores were in some respects 400 years ahead of their time. Two centuries later Robert Hooke suggested that fossils were evidence of earlier lifeforms, just as archaeological artefacts were evidence of former empires.

Nicolas Steno was Professor of Anatomy at the University of Padua when in 1667 someone sent him a shark's head for dissection. He noticed the similarity between the shark's teeth and the so-called tongue-stones which were often found in certain rock formations. As we now know, these were really fossilized sharks' teeth.

Experiments with heat by Steno and an earlier enquirer, the Neapolitan physician Fabio Colonna, proved that fossils were organic in origin. Steno argued, as Hooke had done, that the change in composition between living and fossilized sharks' teeth was the result of infiltration by mineral particles, a process we now call permineralization.

Steno now went further, asking how one solid object, a fossil, came to be inside another, the rock; and not only fossils but crystals, mineral veins, even whole layers of one kind of rock within or across another. Two years later, in 1669, he published his conclusions as *De solido intra solidum naturaliter contento dissertationis prodromus*

("Preliminary discourse to a dissertation on a solid body naturally contained within a solid").

*Dissertationis prodromus* was a remarkable leap of insight into the geological formation of the Earth's surface. It contained four defining principles of the future science of stratigraphy, which profoundly influenced Scottish scientist James Hutton, the founder of modern geology. They state that:

- when a stratum was being formed, it must have been on top of a harder substance which prevented it from sinking further.
- when the upper stratum was being formed, the lower stratum was already solid.
- when any given stratum was being formed it was either contained on its sides by another solid substance, or it spread across the entire surface of the Earth; so if you can see the bared sides of the stratum, either a continuation of the same stratum must exist, or another solid substance must be found which stopped the stratum from spreading.
- if a body of material cuts across a stratum, it must have formed after that stratum.

*Dissertationis prodromus* was Steno's last major work. Profoundly religious, he converted from his native Lutheranism to Roman Catholicism in 1667 and eight years later was ordained as a priest. Geology's loss was Catholicism's gain; he was in time appointed auxiliary Bishop of Münster, in which role he was mocked by Protestants for his pious support of the poor. He died in poverty of malnutrition and was regarded locally as a saint. Pope John Paul II canonized him in 1988.

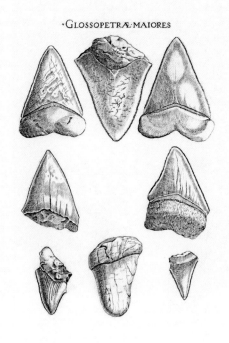

·GLOSSOPETRÆ·MAIORES

*LEFT: Steno's leap of imagination came when he compared the teeth of a dissected shark and its skull to glossopetrae – known as "tongue stones" – that were found in certain sedimentary rocks. Ancient scholars, such as Pliny the Elder, had suggested that these stones fell from the sky or from the Moon.*
*BELOW: The frontispiece for Steno's landmark publication.*

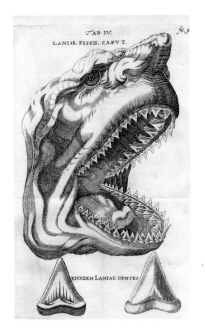

TAB. IV.
LAMIAE. PISCIS. CAPVT.

EIVSDEM LAMIAE DENTES

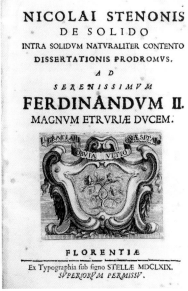

NICOLAI STENONIS
DE SOLIDO
INTRA SOLIDVM NATVRALITER CONTENTO
DISSERTATIONIS PRODROMVS,
AD
SERENISSIMVM
FERDINANDVM II.
MAGNVM ETRVRIÆ DVCEM.

FLORENTIÆ
Ex Typographia sub signo STELLÆ MDCLXIX.
SVPERIORVM PERMISSV.

# Antonie van Leeuwenhoek

## (1632–1723)

## Bacteria

Antonie van Leeuwenhoek lived his whole life in the Dutch city of Delft. He was not a scientist, but through his thorough scientific method and powers of observation, he made many discoveries that have earned him a reputation as the first microbiologist.

Perhaps there was something in the Delft air that gave its citizens an eye for detail. Van Leeuwenhoek, observer of micro-organisms, was a contemporary of Dutch fine artist Johannes Vermeer. They were born only a few days apart, and van Leeuwenhoek acted as Vermeer's executor after the painter's death in 1675.

Delft, famous today for its blue pottery, was also a centre of printing and tapestry weaving. Van Leeuwenhoek himself opened a draper's shop in 1654, the year of the Delft Thunderclap, an explosion in a gunpowder store that destroyed much of the town and left thousands injured. He and his shop survived.

One essential tool of a linen merchant was a magnifying glass, which traders used to count the threads and ensure consistent quality from their weavers. Van Leeuwenhoek was fascinated by magnification and taught himself to make his own bespoke lenses. He became adept and is believed to have produced examples with a magnification of up to 500 times, in an age when ordinary magnifying glasses, and the first rudimentary microscopes, were capable of only 30 or 40.

He began to study the magnified natural world, making sketches of parts of bees, lice and fungus, which impressed a Delft anatomist, Reinier de Graaf. At de Graaf's insistence he wrote with his observations to the Royal Society in London, starting a correspondence which he continued throughout his lifetime, amounting to nearly 200 letters.

The Royal Society was impressed by van Leeuwenhoek's work and he felt encouraged to continue, examining ever-smaller manifestations in nature. He is believed to have been the first to observe protozoa, single-celled organisms, in 1674; and two years later he discovered the existence of bacteria in water from pond, puddles and wells, and in the human mouth and gut.

Despite the reputation that he had established with the Royal Society by then, van Leeuwenhoek's bacteria were met with scepticism when he reported his new findings to the Society. Microscopy was in its infancy and the existence of bacteria had never even been suspected. Water might be muddy, but surely not teeming with invisible life? Van Leeuwenhoek held his ground however, and finally the Society dispatched a team of observers to verify his claims. He was elected a member of the Society in 1680.

Antonie van Leeuwenhoek was a pioneering explorer of our microscopic world. Robert Hooke, a contemporary fellow explorer, acknowledged his leadership in the field. Van Leeuwenhoek also discovered spermatozoa and, with his friend Reinier de Graaf, transformed our view of the process of reproduction in insects and humans. His discovery of bacteria made our modern understanding of disease and hygiene possible.

# Daniel Gabriel Fahrenheit

## (1686–1736)

## Fahrenheit temperature scale

Best known for his temperature scale, Daniel Fahrenheit introduced new standards of precision in his work as an instrument maker. In his pursuit of thermometer perfection he made important discoveries about melting and boiling points.

Fahrenheit was born into a wealthy merchant family in Gdansk. His parents died suddenly (after eating poisonous mushrooms) when he was sixteen and his new guardian packed him off to Amsterdam to acquire business skills. In the cosmopolitan Dutch city, he first became fascinated by scientific instruments and the new science that prompted them.

After he came of age he travelled throughout northern Europe meeting other instrument makers and finding out what scientists wanted from them. Early in his journey, Fahrenheit watched the Danish thermometer maker Ole Rømer at work in Copenhagen. The visit, in 1708, made a lasting impression on him and he began to make his own thermometers soon afterwards – he is credited with making the first alcohol-driven thermometer in 1709 and the first mercury instrument in 1714. In 1717 he returned to the Netherlands and set up shop as a glassblower, making the fine glass tubes required for thermometers and barometers.

In order to make his instruments as accurate as possible he studied the properties of all their components – the way in which alcohol, mercury and glass expanded and contracted with temperature for example. He also made important discoveries about the behaviour of water. The freezing and boiling points of water were often, although not exclusively, used as fixed points by thermometer makers; and Fahrenheit discovered not only that water can remain in liquid form even below its supposed freezing point but that the boiling point of water and other liquids varies considerably with atmospheric pressure and altitude.

Thermometers had been made for about a hundred years before Fahrenheit joined the profession; but there was no standard scale. Instrument makers devised their own, based on little more than "as cold as it can be" through "comfortably warm" to "boiling" and beyond. Any divisions between such "fixed" points as freezing and boiling were purely arbitrary. When he visited Rømer, Fahrenheit made a note of Rømer's calibration of his thermometers. Fahrenheit saw the Dane mark the temperature of iced water, and the temperature of water that was at "body temperature". Measuring the distance between the two points he then added half that distance below the ice point, to allow room for temperatures below freezing; then he divided the whole length into a scale of 22.5 units or degrees. So the lowest point was 0°, freezing point 7.5° and body temperature 22.5°.

When Fahrenheit came to mark his own thermometers up, he thought he was following Rømer's practice. But Rømer's idea of body temperature was water that was comfortable to the touch, not (as Fahrenheit assumed) the actual temperature of blood. Fahrenheit fixed his 22.5 "body temperature" mark by putting a thermometer in his mouth. For greater precision he also divided his degrees by four – so 22.5° became 90° and freezing point, Rømer's 7.5°, became 30°. Later he arbitrarily changed 90° to 96° and 30° to 32° to make certain calculations easier. According to this scale, he declared, water boiled at 212°.

The Fahrenheit scale, arrived at by such random decisions by both Rømer and Daniel, was widely adopted in both the Netherlands and Britain, and carried by immigrants from those countries to the New World of North America, where it remains the standard temperature scale. By the time it emerged that Fahrenheit's boiling point was out by a few degrees it was too late. The scale was quietly adjusted so that body temperature was the one used today, 96.8°F. Water still freezes at 32°F.

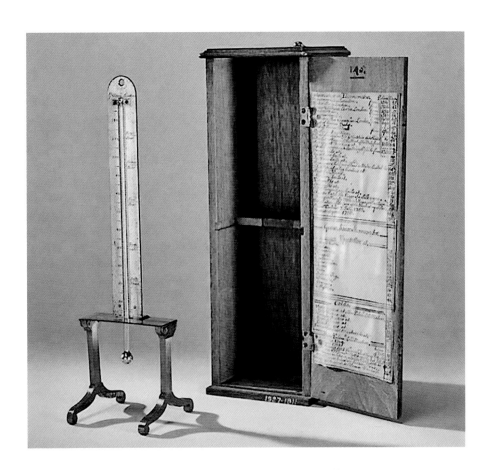

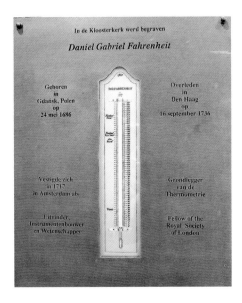

ABOVE: *A mercury thermometer graduated from -20 to 210 Fahrenheit, used for chemical experimentation. It was made by George Adams of Fleet Street, instrument maker to King George III.*

LEFT: *The thermometer on Daniel Fahrenheit's tomb. Today, the United States is the only major industrialized nation to use the scale.*

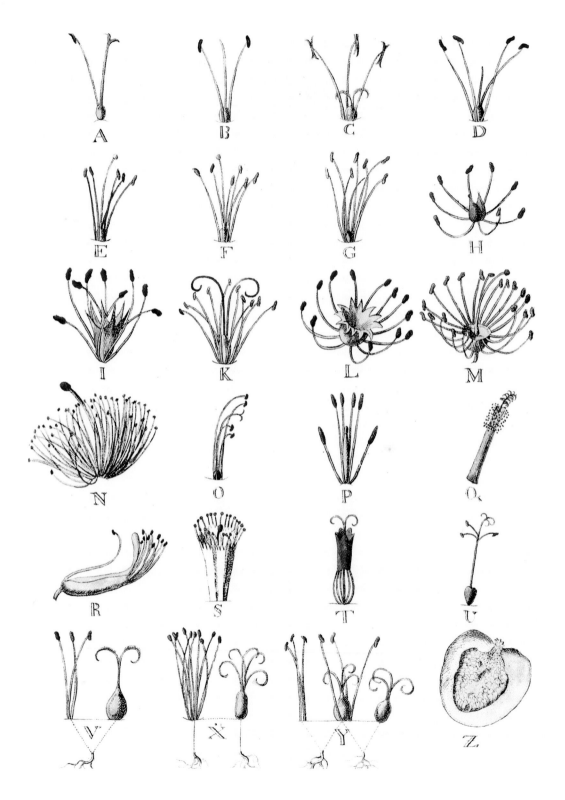

# Carl Linnaeus

## (1707–1778)

## A system of classification of the natural world

From his early childhood, when his parents would give him flowers when he was ill or sad, Carl Linnaeus was fascinated by botany. Even his surname was chosen (by his father) because of a tree growing on the family plot – Linnaeus is the Latin for a lime tree.

Carl's father and grandfather before him were all rectors of the local Protestant church where he grew up in Sweden. His teachers noticed his interest in botany and encouraged it; and to the great disappointment of his mother, Carl chose not to follow his family into religious service but to study botany and medicine at Uppsala University.

Any scientist's basic response is to make order out of chaos, to find patterns and laws. Linnaeus had been taught a botanical classification devised by Joseph Pitton de Tournefort. That system, although rather arbitrary and artificial, was the first to differentiate between genus and species. It was during a field trip to Lapland that Linnaeus first thought about classifying species according to inherited characteristics, when the dental pattern of a horse's skull caught his eye.

He first published his ideas in 1735 in a short pamphlet called *Systema Naturae* ("The System of Nature"), introducing a hierarchy based on classes, orders, genera, species, and varieties. Linnaeus's initial approach was based on the reproductive systems of plants – he described a plant's stamens and pistils as its husbands and wives. In subsequent expansions of his system, however, he took greater note of other characteristics such as flowers and fruit and he considered his masterpiece to be his 1737 book *Genera Plantarum* ("Genera of Plants").

The logic of Linnaeus's classification distinguished it from other systems and it was quickly adopted by other botanists. As the new method spread, plant collectors from all over the world sent him descriptions and samples, with which he was able to confirm and expand his approach. In his lifetime he produced twelve editions of *Systema Naturae* and six of *Genera Plantarum* as well as numerous other works.

The success of Linnaean taxonomy standardized the naming of all living things. It transformed the sciences of biology and botany and gave practitioners a common language with which to refer to their subjects – binomial nomenclature. Before it, the same flower might have rejoiced under any number of common names depending where it grew, what it was used for, or which plant collector had discovered it there. Now everyone could use the same name, which could be reduced to just two words, the genus and the species.

For example, in an ordinary conversation between biologists about their pets, there is no need to reel off the dog family's full Linnaean classification (including some layers added since Linnaeus's time): Animalia (the Kingdom, to distinguish it from Plantae); Chordata (the Phylum); Mammalia (class); Carnivora (order); Canidae (family); Caninae (subfamily); Canini (tribe); Canina (subtribe); and *Canis* (genus). *Canis* refers to all the members of a particular branch of the dog family. These include *Canis lupus* (the grey wolf); *Canis latrans* (the coyote); a number of other distinct species of wolf and jackal; and *Canis familiaris*, the common domestic pet dog and all its subspecies.

Starting with Linnaeus, all living things are assigned a discoverer, the person who first recorded them under Linnaean classification. Naturally many of the earliest records are credited to Linnaeus himself, including our own genus and species, forever recorded as *Homo sapiens* (*Linnaeus*, 1758).

*LEFT: Linnaeus classified plants on the basis of their reproductive parts, as illustrated here in 1736 by Georg Ehret, one of the most esteemed botanical illustrators of the time. Ehret met Linnaeus in the mid-1730s when Ehret was painting the plant collection of a wealthy banker.*

# Daniel Bernoulli

## (1700–1782)

## The Bernoulli Principle

The Bernoulli family were a remarkable dynasty of mathematicians: the brothers Jacob and Johann, and Johann's sons Nicolaus, Daniel and Johann II. Daniel was considered the most brilliant of the younger generation, and his research into the pressure of fluids helped the modern world to fly.

There was considerable brotherly rivalry between Jacob and Johann the elder, and when Jacob was appointed professor of mathematics at the University of Basel, Johann was extremely jealous. But the brothers did important work together on early applications of calculus, which had recently been discovered independently by Sir Isaac Newton and Gottfried Leibniz.

When Jacob died in 1705, Johann triumphantly took over his chair at Basel, and transferred his jealousy from his brother Jacob to his son Daniel. It is said that in later life, when father and son entered the same scientific competition and came equal first, the father's embarrassment at not outdoing his son was so great that he threw Daniel out of the family home.

Johann the elder's father had tried to steer him into a business career, something Johann resisted in favour of studying medicine. Despite that experience with his own father, Johann demanded that his son Daniel study business. Daniel also resisted, and he too chose medicine, but only on the condition that his father teach him mathematics in his spare time. Another of Johann's students was Leonhard Euler, and Leonhard and Daniel were great friends.

Euler and Bernoulli worked together on the relationship between blood pressure and the speed of blood flow. Daniel's particular interest was the theory of

conservation of energy – he noted the principle of a change from kinetic to potential energy in a moving body which accompanied a gain in height, and applied it to fluids, in which kinetic energy was exchanged for pressure.

Daniel Bernoulli published his magnum opus, *Hydrodynamica*, in 1738. The name was a word of his own invention and was adopted for the new field of engineering, hydrodynamics, which the book opened up. In it he took the conservation of energy as his starting point and considered the efficiency of hydraulic machines. It contains the first exposition of the kinetic theory of gases.

Above all it includes Bernoulli's Principle, the rule that an increase in the speed of a fluid goes hand in hand with a decrease in pressure or a decrease in the fluid's potential energy. Leonhard Euler devised the equation that goes with it. Besides its application to hydrodynamic engineering, Bernoulli's Principle is central to aerodynamics. It is the reason that aeroplane wings have their distinctive cross-section, which encourages lift and flight.

Daniel's father was so jealous of *Hydrodynamica* that he plagiarized it in his own 1739 book, *Hydraulica*, which he backdated to 1732 so that it appeared that he, Johann Bernoulli, had thought of Bernoulli's Principle first. Johann harboured this resentment of his son's success until the day he died.

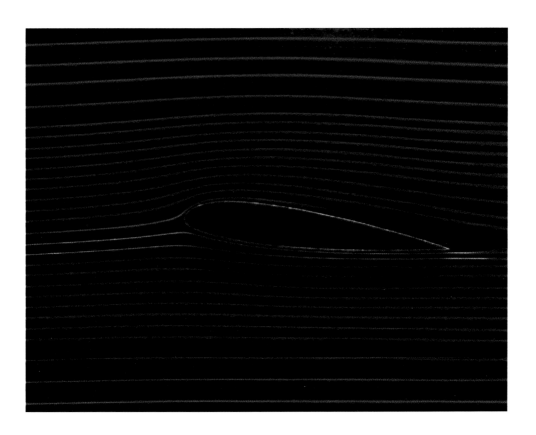

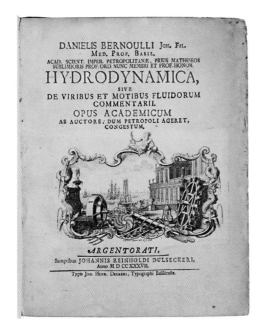

*ABOVE: A pattern of air flow demonstrating the Bernoulli Principle, which states that the internal pressure of a gas is lowered the faster it travels. This principle gives an aeroplane wing lift.*

*LEFT: Bernoulli's great work was shamelessly plagiarized by his father.*

*OPPOSITE: Daniel Bernoulli was actually born in the Netherlands, then under Spanish rule. The talented family of mathematicians moved to Basel in Switzerland to escape Spanish persecution of Protestants.*

# Anders Celsius

## (1701–1744)

## Celsius temperature scale

Science was in Anders Celsius's blood. His father was an astronomer; his grandfathers were an astronomer and a mathematician. But when he first devised his famous temperature scale, he got it upside down.

Anders Celsius followed his father as professor of astronomy at the University of Uppsala in Sweden. The early years of his career were distinguished by studies of the phenomenon of the aurora borealis, the Northern Lights. He published a large volume of papers on the subject and was the first to observe the effect of heightened auroral activity on the Earth's magnetic field. Celsius erected Sweden's first purpose-built astronomical observatory on the roof of his house in Uppsala.

Uppsala has an ancient tradition of learning. Its university is the oldest surviving institution of learning in Scandinavia, founded in 1477 and offering classes in astronomy from as early as the 1480s. Its chair of astronomy was established in 1593. The city is home to the Royal Society of Sciences in Uppsala, Sweden's oldest royal academy. It was founded in 1710 and Celsius was its secretary from 1725 until his death. Carl Linnaeus, the botanist who devised the modern classification of all living things, was a fellow member.

It was in a paper delivered to the Royal Society that Celsius first proposed his new temperature scale. Thermometers had been around for over a century, but their scales were often arbitrarily decided by their individual manufacturers. Ole Rømer in Copenhagen and Daniel Fahrenheit in Amsterdam were among those attempting to introduce a more rigorous, reliable standard measurement. Fahrenheit's found widespread use in Britain and Holland, while another, devised by Frenchman René de Réaumur, was popular in France, Germany and Russia. It remains in use in the cheese industries of Switzerland and Italy, and the confectionery industry in the Netherlands.

There were 180°F (Fahrenheit) between the freezing and boiling points of water, and 80°Re (Réaumur). For Fahrenheit, water froze at 32°F, in the belief that brine froze at the lowest possible temperature, 0°F. Réaumur set the freezing point of water at 0°Re and may have been the first to allow for negative temperatures.

There were problems with both scales. Although Fahrenheit had demonstrated variations in the boiling temperatures of water and other liquids depending on air pressure, neither he nor Réaumur had set defining conditions of altitude or pressure for their measurements. Both Réaumur's 80° of difference and Fahrenheit's 180° (derived from Rømer's 60°) were based on incorrect understandings of the possible extremes of temperature.

Celsius proposed a rational new system in which everyday experience – from freezing to boiling – was measured across 100°. Crucially, he also set standard conditions for his scale: temperature readings were to be calibrated at sea level with an air pressure of 760mm ("one atmosphere") and a mercury thermometer. Fahrenheit had invented a mercury thermometer, but both he and Réaumur also used less reliable alcohol-based devices.

These conditions made the Celsius scale altogether more accurate, so his proposal was widely adopted. Known for two hundred years simply as the centigrade ("hundred-step") scale, it was officially renamed in Celsius's honour in 1948. Celsius, however, went a step too far in his efforts to improve on his predecessors; he set the temperature of freezing water at 100°C and its boiling point at 0°C. His friend and colleague Linnaeus quietly reversed them a year after Celsius's death.

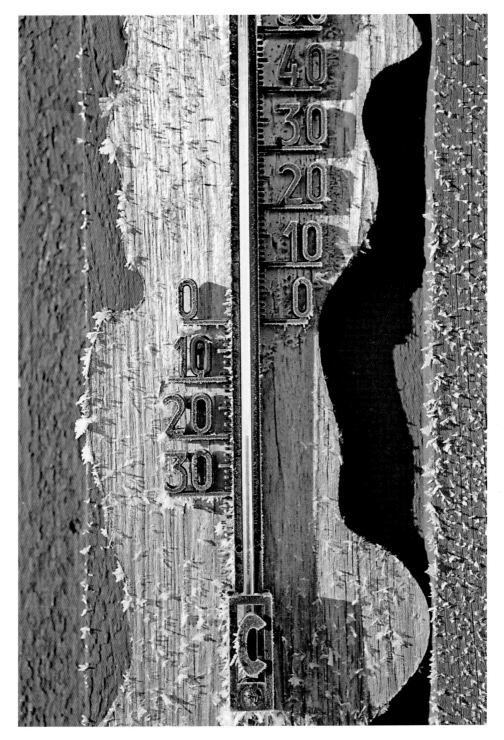

*ABOVE: An external thermometer from Anders Celsius's native Sweden.*

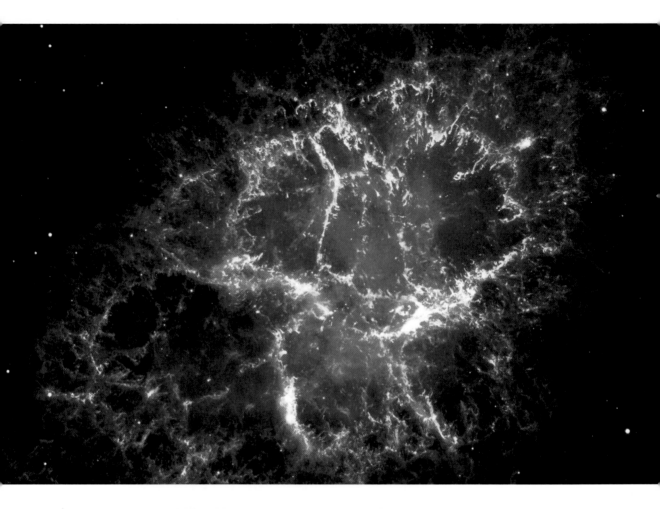

*ABOVE:* The Crab Nebula, No.1 or M1 in the Messier Catalogue. Messier came across it in 1758 while looking for comets. This image was captured by the Hubble Space Telescope in 2013.

*LEFT:* Messier 101, the Pinwheel Galaxy.

*OPPOSITE:* Messier's observatory was located on top of a tower at the Hôtel de Cluny in Paris.

# Charles Messier

## (1730–1817)

## A catalogue of astronomical objects

Nebulae, galaxies, stars, planets, moons, asteroids, comets: space is full of objects. Some of them are fixed in their positions; others are on the move. For those exploring the night sky by telescope it is helpful to know which is which. French astronomer Charles Messier produced a handy guide in 1774.

Messier's interest in the night sky was first spurred in 1744 by the arrival of a remarkable comet with six tails. One of the brightest ever recorded, it was visible for seventy-one days as it sailed across the universe. Messier was thirteen at the time; Catherine the Great, fourteen and on her way from Prussia to Russia to be wed, also witnessed the auspicious event.

At the age of twenty, Messier was engaged as an assistant astronomer in the French Navy, which still used stars for navigation. There he learned the importance of recording his observations both at naval observatories and from his own base in Paris's Hôtel de Cluny.

Messier's particular passion was, naturally enough, comets. In the course of his astronomical life he discovered thirteen of them. To do so he had to be able to distinguish between fixed objects and those that had moved, however imperceptibly, from one observation to the next.

To this end he compiled and published a useful catalogue of distant objects on whose position an astronomer could rely. The first edition in 1744 contained forty-five such things – galaxies, nebulae and star clusters – including seventeen which he and his assistant Pierre Méchain had discovered themselves. The final edition, published in 1781, contained 103, to which astronomers have added seven that Messier is known to have observed.

Messier's *Catalogue des Nébuleuses et des Amas d'Étoiles* ("Catalogue of Nebulae and Star Clusters") has some limitations. The pamphlet is somewhat chaotically organized – his items are not listed either by type or by location. He made his observations through a refracting telescope only 100mm in diameter and they are therefore limited to what is visible through such

an instrument; and he made them from the Hôtel de Cluny, so they are restricted to objects in the night sky over Paris.

Modern telescopes are far more powerful, and located all over the world; but Messier's contribution to early space exploration is so great that the objects which he catalogued are still known as M1 to M110. For amateur astronomers scanning the heavens with only the modest telescopes at their disposal Messier's list of some of the brighter objects in the night sky is invaluable even today.

# Joseph Black

## (1728–1799)

## Carbon dioxide

By the time Joseph Black was conducting his chemical experiments, matter was deemed to exist in one of five principal states: water, fire, earth, metal and salt. Black's fascination with what happened when those principals were combined in various ways led to some important discoveries.

Joseph Black was born in Bordeaux, the son of Scottish and Irish parents working in the wine industry. But instead of following them into the same line of work Black studied at the universities of Glasgow and Edinburgh in Scotland, with such distinction that in later life he was appointed Professor of Medicine, first of the former and then of the latter.

The study of medicine required a knowledge of chemistry, as it applied to treatment and cure. For his thesis at Edinburgh Black made a particular study of magnesium carbonate, and after his graduation he continued to experiment with the substance and with calcium carbonate (chalk). He demonstrated that when he heated each substance, the residue (magnesium oxide or calcium oxide) was lighter than the original material. His conclusion was that the weight must have been lost as a gas, which he called "fixed air". It's better known today as carbon dioxide. Adding a solution of potassium carbonate to the residue restored it to its original condition and weight.

Before Black's discovery, gases were not thought to be distinct chemical compounds, just forms of air with different levels of purity. Black proved that his "fixed air" was not merely air by demonstrating that it could not support animal or vegetable life. The discovery of carbon dioxide was a significant moment for chemistry and in the second half of the eighteenth century other chemists identified other common gases – oxygen, nitrogen and hydrogen.

Black made a second important discovery during experiments with fire and the three principal states of water – ice, water and steam. He found that applying heat to ice at precisely its melting point did not raise its temperature; it simply produced more water. Similarly, at boiling point, the temperature of water does not rise; it simply produces more steam.

Since water turns to steam at a given temperature, one would expect all the water to turn to steam at that temperature. In fact the temperature of many substances falls at melting and boiling points. Black deduced that the heat which he applied was absorbed as energy used, for example, in the conversion of ice to water. The heat, he decided, was thus "hidden" in the ice. He called it latent heat, from the Latin word for lying hidden.

With the discovery of latent heat, Joseph Black single-handedly launched the science of thermodynamics. It had the immediate effect of spurring on the industrial revolution. James Watt, who made scientific instruments for Glasgow University, consulted Joseph Black about his work. He went on to design the two-cylinder steam engine, which transformed the world's mining, milling and manufacturing industries in the following century. Watt's improvement of Thomas Newcomen's original steam engine essentially took advantage of Black's phenomenon of latent heat.

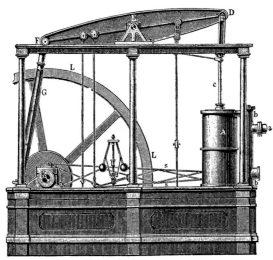

ABOVE: *A block of magnesite, magnesium carbonate.*

LEFT: *A James Watt steam engine from 1776. Watt had built experimental equipment for Joseph Black at Glasgow University. Realizing the energy losses of latent heat from Thomas Newcomen's steam engine, he improved it by adding a condenser.*

OPPOSITE: *An engraved portrait of Joseph Black.*

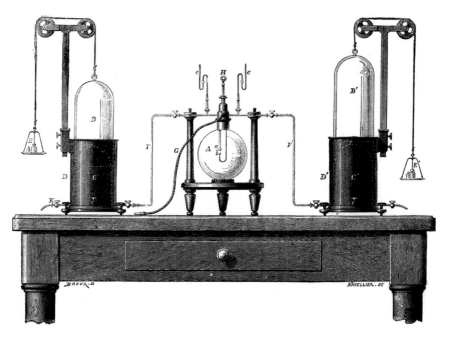

TOP: *A portrait of Antoine-Laurent Lavoisier and his wife and collaborator Marie-Anne Pierrette Paulze,*
*commissioned from the French painter Jacques-Louis David in 1788.*
ABOVE: *Antoine Lavoisier's apparatus for synthesizing water from hydrogen (left) and oxygen (right).*

# Antoine Lavoisier

## (1743–1794)

## Law of conservation of gases

Joseph Black's discovery of carbon dioxide sparked a wave of chemical experiments to find and identify other gases. French chemist Antoine Lavoisier was interested in the role of the air in the combustion of mineral salts and the production of rust. In the process he gave oxygen its name.

At the end of the seventeenth century a new theory about combustion was gaining ground in the scientific community. Much favoured and promoted by the German chemist Georg Ernst Stahl, it held that a previously unrecognized "fiery substance" called phlogiston lay within other materials and burnt off when it was ignited. If the remains were then reheated with another substance containing phlogiston, such as charcoal, they would reabsorb the phlogiston.

According to Stahl, soot was almost pure phlogiston, and wood was a compound of phlogiston and ash. Metal was released from metallic minerals by heating them with phlogiston. But although the existence of this magical substance explained many observable phenomena, it could not account, for example, for an increase in weight during the combustion of tin and lead. Should they not be lighter after the phlogiston had been released?

Antoine Lavoisier was a wealthy French nobleman who trained as a lawyer but excelled in mathematics and was an enthusiastic experimenter in the emerging discipline of chemistry. With his mathematical skills he helped transform chemistry from a qualitative to a quantitative science. Instead of simply noting that certain combinations produced certain effects, for example, he wanted more precise answers about how much of each reagent was required and how much residue remained.

Lavoisier believed that in any chemical reaction, as he put it, "Nothing is lost, nothing is created, everything is transformed." A loss of weight on one side of an equation must imply a gain on the other; and if that gain was not found in solid matter, then it must be fluid – a gas. In the wake of English chemist Joseph Priestley's experimental isolation of "dephlogisticated air", Lavoisier repeated the experiment and gave the gas a name – oxygen.

The existence of oxygen confirmed Lavoisier's theory that matter cannot be created or destroyed, but it can change forms; and this became Lavoisier's Law of the Conservation of Mass. In some countries it is known as Lamonosov's Law, after the Russian scientist who reached the same conclusion independently and concurrently.

Lavoisier made many other contributions to science. By isolating oxygen he was also able to identify nitrogen, the other major component of air. He standardized chemical nomenclature in the form we know today. He chaired the committee that designed the metric system of measurement. He made studies of air pollution, water purification and the practicalities of urban street lighting.

He oversaw the reliable supply of gunpowder to French governments before and after the French Revolution and funded a printing press, which produced the revolutionary newspaper *La Republicain*. Unusually for a nobleman, he was committed to public service and established two organizations – the Lycée and the Musée des Arts et Métiers – to provide scientific education for the public.

Antoine Lavoisier's wealthy background allowed him to indulge in his passion for chemistry. But it was his undoing when it made him an enemy of the people during the French Revolution. One of his lucrative roles had been as a tax collector for the Ancien Régime. "It took them only an instant to cut off that head," said a colleague, "and a hundred years may not produce another like it."

# William Herschel

## (1738–1822)

## Uranus

It's not every day that someone discovers a new planet. Uranus was the first new planet to be observed since antiquity. William Herschel, who discovered it, was an almost entirely self-taught astronomer. He came to Britain from his native Hanover not to gaze at the stars but to play the oboe in a military band.

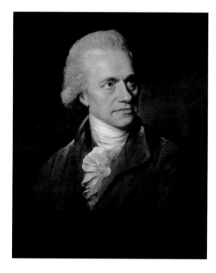

Like his father before him, William Herschel was an oboist in the band of the Hanover Guards, his home state's regiment. The Duke of Hanover was, by the mysteries of genealogy, also the King of Great Britain, George II; and the regiment was posted to England. After leaving his regiment William made a living there as musician and composer of twenty-four symphonies, eventually settling in the spa town of Bath. There he was joined by his brothers and sister, all accomplished musicians, and they often performed together.

In a bid to improve his social standing in Bath, Herschel read avidly in many aspects of natural philosophy, or science, as it was becoming known. Beginning with the science of music he also read a beginners' guide to Newtonian astronomy called *Astronomy Explained upon Sir Isaac Newton's Principles and Made Easy to those who have not Studied Mathematics.* This led him to learn about optics and, after taking lessons from a mirror maker, he constructed his own reflecting telescope in 1773 with the help of his siblings. Six years later he was accepted into the prestigious Bath Philosophical Society.

Herschel's initial interest was in binary stars and he discovered more than 800 of these star pairs using the telescope in his Bath back garden. It was during his comprehensive search of the skies that he stumbled upon an unidentified disc. He thought it might be a comet somewhere beyond the orbit of Saturn, but after comparing notes with other astronomers it was confirmed as the first new planet to be discovered since ancient Babylonian astronomers had identified Saturn and Jupiter. Herschel named it after King George III and in countries such as France, with no love of Britain or Germany, it was known as Herschel. Eventually astronomers agreed on the name Uranus, after the Greek god of the skies.

William Herschel's discoveries did not end there. In the course of his observations he identified 2,500 new star clusters and nebulae, and found two new moons each for Uranus and Jupiter. In addition he identified infrared light, and is credited with coining the word "asteroid". His telescopes were in great demand all over Europe and he is known to have sold over sixty of them, including one to the King of Spain.

Much of his work was completed with the help of his sister Caroline, who earned a reputation as an astronomer in her own right. The house in Bath that she and William shared until William's marriage is now the Herschel Museum of Astronomy.

*ABOVE RIGHT: Uranus is the seventh planet from the Sun and the third-largest in the Solar System. It is a giant planet with twenty-seven known satellites.*
*RIGHT: The observatory from which Herschel made his 1781 discovery.*
*ABOVE: A portrait of the king's astronomer.*

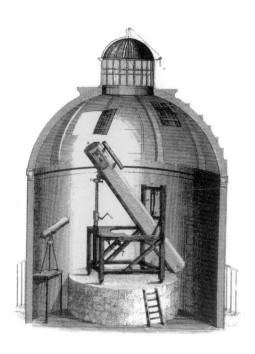

*RIGHT AND BELOW: Two photographs from the Herschel Museum in Bath: the Herschel telescope and a room set as it would appear in the Georgian period. Both brother and sister were gifted musicians.*

# Caroline Herschel

## (1750–1848)

## Comets

She discovered five comets. She was the first woman scientist to be paid for her astronomical work, and she was the first woman to receive membership of the Royal Society, Britain's national academy of the sciences. Today's astronomers, male and female, owe a debt of gratitude and admiration to Caroline Herschel.

Raised by an illiterate mother who denied her an education because of her gender, Caroline Herschel was occasionally taught secretly by her father. When he died she was finally able to indulge her thirst for learning when she left her native Germany to join her brother William in Bath, England.

William was an accomplished musician and Caroline learned to sing so that she could join him in his performances. She also acted as his personal assistant, running his household and, when he became interested in astronomy, recording his observations. She proved to have a good understanding of William's work and he soon asked her to do the repetitive work of searching the skies for comets, a strip at a time. Although she found the work tedious at first, eventually her interest was piqued.

Her first discovery was not a comet but a new nebula, an event which encouraged William to find nebulae of his own. Frustrated with the quality of his equipment he began to make his own telescopes, and Caroline assisted in the polishing of mirrors and the grinding of lenses. William made her one of her own in 1783, with which he asked her to help him search the skies.

Before the end of the century Caroline discovered at least five new comets. Five are credited entirely to her; two others were discovered simultaneously by

other observers; and an eighth was the rediscovery in 1795 of a comet which had first been observed in 1786. She was not the first woman to discover a comet – that honour goes to Maria Kirch, whose early-eighteenth-century observations were at the time credited to her husband.

After her brother was appointed Astronomer Royal in 1782, it was as his assistant that Caroline became the first woman to receive a royal stipend, and thus the first professional female astronomer. She catalogued her and William's discoveries, adding over 500 objects to the 2,000 stars, clusters and nebulae previously known. Although it was published under William's name, her achievements were widely recognized, and in 1828 she was awarded the Royal Astronomical Society's Gold Medal and, seven years later, its honorary membership.

She outlived her brother by twenty-six years and died in 1848 at the age of 97. After her brother's death she returned to Germany and continued her work. On her ninety-sixth birthday the King of Prussia awarded her a Gold Medal for Science.

*LEFT: Caroline Herschel pictured working with her brother, Sir William Herschel. She was not only the first woman member of the Royal Society, she was also Britain's first professional female scientist.*

# Jacques Charles

## (1746–1823)

## Law of ideal gases

Scientists are often jealously protective of their claims of priority in matters of discovery. So it comes as a surprise that Charles's Law of Ideal Gases was named not by Jacques Charles himself but by fellow Frenchman Joseph Louis Gay-Lussac, who also has two gas laws named after him.

Jacques Charles is best remembered today for his hunch that hydrogen would make a good source of lift for balloons in the race to complete the first ever manned flight. As a physicist he reached this conclusion after reading the works of fellow gas investigator Robert Boyle; and he put it to the test with dramatic results.

Working with two engineering brothers, Anne-Jean and Nicolas-Louis Robert, Charles launched the world's first unmanned hydrogen balloon in August 1783, just eight months after the unmanned hot-air balloon of the Montgolfier brothers. Hydrogen was supplied by pouring a quarter of a ton of sulphuric acid onto half a ton of rusted scrap iron. In the village where it landed, locals were so alarmed that they attacked the balloon with pitchforks: the future had arrived. In December of that year Charles and Nicolas-Louis Robert became the first men to fly in a hydrogen balloon, travelling 43km (27 miles) just ten days after Pilâtre de Rozier and Marquis d'Arlandes had made the first-ever human flight, in a Montgolfier balloon. Benjamin Franklin, who had helped to fund Charles's balloon, was among the onlookers for its inaugural flight. Hot-air balloons became known as Montgolfières and hydrogen balloons as Charlières.

The Robert brothers and Charles next attempted to build a balloon that could be steered – a dirigible. But the oars and rudder with which they planned to do this were ineffective in the air. Indeed, the Roberts were powerless to prevent the balloon rising to dangerous heights until their passenger, the Duke of Chartres, used a dagger to puncture some of the hydrogen cells. Charles was persuaded as a result to invent the hydrogen valve.

Although Charles never flew again after that first day, he remained interested in the behaviour of gases. In one experiment he filled five identical balloons with five different gases, to observe the change in volume of each when heat was applied. He heated them all to 80°C and saw that they had all expanded by the same amount.

Charles did not publish this observation, but when Gay-Lussac was looking into the relationship between gas and temperature he generously gave Charles the credit for first observing it. Charles's Law of Ideal Gases states that the volume of a gas at constant pressure increases linearly with the temperature of the gas. The gases in question are described as ideal because the law doesn't take into account any other factors such as the interaction of particles within the gas. The ideal can therefore be compared against the practical reality of a given situation.

Of Gay-Lussac's own two Laws of Gases, one is genuinely his and another was actually discovered earlier by someone else, unknown to him. The former is Gay-Lussac's Law of Combining Volumes, which states that when gases react together the volume of the resultant gas is in a simple whole number ratio to the volume of the original gases. The latter, that the pressure of a gas of fixed mass and fixed volume is directly proportional to the gas's absolute temperature, can be seen as a corollary of Charles's Law, and was first stated by another Frenchman, Guillaume Amontons. It's more usually known now as the Pressure Law of Gases or Amontons' Law. Gay-Lussac's best-known contribution to science is the discovery of the composition of water molecules – two atoms of hydrogen and one of oxygen, $H_2O$.

*ABOVE: An illustration of the hyydrogen-filled balloon (or machine aerostatique) of Jacques Charles and the Robert brothers, which took off from the Jardin du Palais de Thuilleries in Paris on 1 December 1783.*

DEUTSCHE BUNDESPOST

40

GAUSSSCHE ZAHLENEBENE

CARL F. GAUSS 1777–1855

*ABOVE: A stamp issued by the German Post Office in 1977 to celebrate the bicentenary of Gauss's birth.*

*RIGHT: In 1796 Gauss showed that a regular polygon can be constructed by compass and straight edge if the number of its sides is the product of a distinct Fermat prime number to a power of two.*

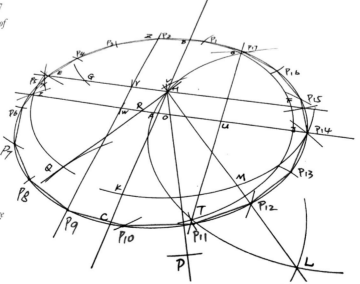
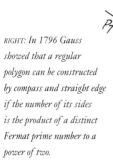

# Carl Friedrich Gauss

## (1777–1855)

## Number theory

*Princeps mathematicorum* – "first among mathematicians" – is a title bestowed on Carl Friedrich Gauss more often than on Newton, Euler or any of the other giants of the discipline. His breakthroughs in number theory guided pure mathematics well into the twentieth century.

Number theory is the branch of pure mathematics which deals with integers – whole numbers – and their relationship to each other. Rational numbers – numbers which can be expressed as fractions using integers – are also covered by number theory. Prime numbers – those only divisible by themselves and the number 1 – are a particular source of fascination by number theorists.

Gauss was a child prodigy. It is said that he could count before he could speak and that as a child he corrected his father's accounts. At primary school he completed an arithmetic problem (the sum of all the integers from 1 to 100) in seconds, which his teacher had hoped would punish the precocious pupil for misbehaviour.

He outstripped all his teachers at university, where he rediscovered the proofs of several theorems without recourse to textbooks. He proved while still a student that it was possible to draw a regular 17-sided polygon using only a straight edge and a compass. Until then it was believed that this could only be done for 3-, 5-, and 15-sided shapes; but Gauss demonstrated it with a heptadecagon and thirty more forms, devising a formula for knowing which polygons were possible by this method.

By the age of 21 he had completed his magnum opus, *Disquisitiones Arithmeticae.* It demonstrated his complete mastery of "arithmetic" (as number theory had been known until then), presenting not only the work of predecessors such as Fermat and Euler but also his own dazzlingly original work. The book became the foundation stone of the modern study of number theory.

Gauss demanded rigorous standards of logic in his proofs. Although *Disquisitiones Arithmeticae* set these standards for all who came after him, much of his best work remained unpublished, for fear of being found illogical and disproved. His diaries contain what modern mathematicians recognize as the roots of theories of L-function, complex multiplication, the Riemann Hypothesis and others. His work on the class number problem was not confirmed until 1986.

But Gauss also made discoveries in other fields. He recognized the possibility of non-Eulerian geometry. Gaussian curvature is the study of landscape forms from a mathematical perspective, developed while he was undertaking a geodetic survey of the Kingdom of Hanover. He proved the General Theorem of Algebra, with regard to complex numbers – three times, in different ways. His studies of light produced ideas still known as Gaussian optics.

Gauss accurately predicted where the newly discovered dwarf planet Ceres might be observed again in 1801. His solution introduced the Gaussian gravitational constant and may have anticipated the Fast Fourier Method, first mooted in 1807 and first published in 1965. His studies of the Earth's magnetic field resulted in the unit of magnetic measurement being called a Gauss.

Advances in all the sciences in the past two centuries mean that we are unlikely ever to see again anyone with such intuitive grasp of so many fields of research. Scottish number theorist Eric Bell (1883–1960) said that if Gauss had actually published all his discoveries he would have advanced mathematics by fifty years.

# Alessandro Volta

## (1745–1827)

## Chemical generation of electricity

Electricity was little more than a novelty of nature before Volta. It occurred as lightning, and as static, known as the amber effect because it was generated by rubbing a piece of amber. It was unpredictable and unmeasurable until Volta discovered a reliable, stable way of producing it.

Alessandro Giuseppe Antonio Anastasio Volta was born in the year in which the Leyden jar was invented, a primitive capacitor which could store an electrical charge and discharge it at will. The device, charged by an electrostatic generator, provided a supply of electricity for experimental purposes. But its output – measured in "jars" – was erratic and it could only be used once, after which it had to be rested or recharged.

Whether by accident of birth or through his Jesuit education, Volta became fascinated by electricity. At the age of 20 he wrote his first scientific paper on the subject of static electricity and at 30 he invented a machine for generating it, the electrophorus. The following year he used it to generate the spark which ignited a gas that he was the first man to isolate – methane. In so doing he demonstrated the principle that lies behind the internal combustion engine, invented 85 years later.

Volta continued to experiment with electricity and invent new devices to control and measure it – a new form of capacitor in 1782, and a series of increasingly accurate electroscopes over the next ten years with which to measure electrical potential and charge. All this resulted in his receiving the highest honour in science, the Royal Society's Copley Medal, in 1794; and all this was before he made his greatest contribution to electrical research.

Volta was far from alone in researching electrical phenomena and in 1791 his countryman Luigi Galvani (the man who gave us the word "galvanization") announced his discovery of "animal electricity". This was the form of electricity, Galvani claimed, which he had observed when a dissected frog's legs twitched upon contact with two different metals.

Volta suspected that animal electricity was the same as any other electricity, and not a distinct force. He proved it by repeating Galvani's experiments, first with a frog and then with an inorganic conductor. The two men entered into a long scientific feud, which only ended with Galvani's death in 1798. Volta was correct: it was not the frog but the use of two different metals which was creating the charge.

Galvani deserves some credit for his accidental discovery of the electric relationship between metals. And horror fans must be grateful for his theory, which reached the ears of British author Mary Shelley and inspired her novel *Frankenstein*. Alessandro Volta pursued the discovery and made one of his own, that the motion of electrons from one metal to another, for example from zinc to copper, through an electrolyte solution such as brine or sulphuric acid, produced a steady current of electricity.

The combination of several of these units of zinc-electrolyte-copper could be piled up to increase the current. The resulting column became known as a Voltaic Pile. It was the first battery, and in France they still used the word "pile" for a battery. For the first time the pile gave researchers a consistent current against which to measure other outcomes of their experiments.

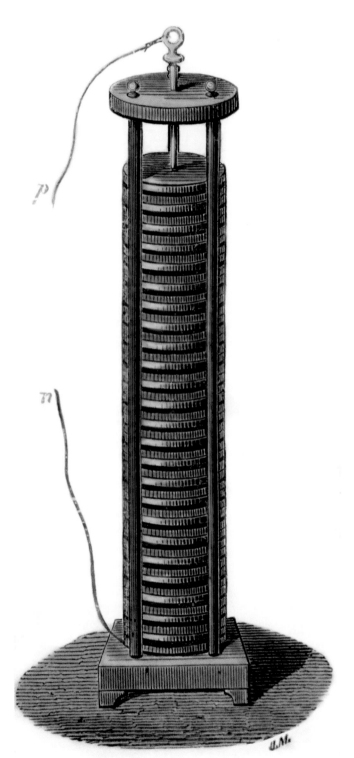

ABOVE: *An engraving showing the charging of a simple Leyden Jar.*
LEFT: *A colour illustration of a Voltaic Pile.*
OPPOSITE: *Volta's name is appropriately attached to an electric vehicle recharging station. Extending battery capacity is the new "space race" of the twenty-first century.*

OPPOSITE: *The famous miners' safety lamp. Ironically, personal safety was not one of Davy's strong suits.*
LEFT: *An 1821 portrait of Sir Humphry Davy.*
BELOW: *The Royal Institution electric battery. Constructed in 1813, this series of copper and zinc plates was the most powerful battery of the time, built by British chemist William Wollaston for Humphry Davy.*

L. GUIGUET

# Humphry Davy
## (1778–1829)
## Anaesthetic effect of nitrous oxide

Humphry Davy, best known for inventing the miners' safety lamp, was not afraid to put himself in danger for the sake of science. He was a compulsive experimental chemist, of whom as a young apprentice his friends said, "He will blow us all into the air."

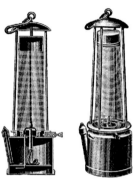

Davy's willingness to experiment on himself nearly cost him his eyesight. Nitrogen trichloride is a highly explosive compound first isolated in 1812 by the French chemist Pierre Louis Dulong, who lost an eye and two fingers to it. Davy's temporary blindness in a laboratory explosion of the material a year later was foolish enough and forced him to hire the young Michael Faraday as his assistant. But then *both* men were injured in a further nitrogen trichloride big bang not long afterwards.

Humphry Davy was no theoretical chemist; his work was all to practical ends. He devised the Davy Safety Lamp in response to another explosion in which nearly a hundred miners lost their lives underground when a naked candle ignited an accumulation of methane. He was consulted by the Royal Navy about protecting the fleet's copper bottoms, the cladding applied to ship to prevent their wooden hulls becoming infested with woodworm. The copper was prone to corrosion, and Davy came up with an electrochemical solution which had the desired effect. Unfortunately it produced a by-product on which seaweed and shellfish fed and created unwelcome drag on the hulls.

Davy was a member of the Pneumatic Institution in the western English seaport of Bristol, a laboratory dedicated to exploring the properties of newly discovered gases. One of them, nitrous oxide, had been synthesized for the first time by the English chemist Joseph Priestley. Davy decided to discover the effect of inhaling it and other gases by doing so himself, in a gas chamber designed for the purpose by his friend, the great Scottish engineer James Watt.

He found that nitrous oxide was intoxicating and that he was prone to giggling under its influence. He named it laughing gas and introduced many of his friends to its effects, including the poet Samuel Taylor Coleridge. Davy became addicted to spending a few minutes in the chamber from time to time. He was less fortunate in his personal experiments with nitric oxide. The gas combined with the air in his mouth to produce nitric acid, which burned the lining of his mouth and throat. He almost died when conducting the same empirical research into carbon monoxide.

Typically, Davy immediately put his mind to nitrous oxide's practical applications, and speculated that it might be useful as an anaesthetic in surgical operations – hitherto surgeons had relied on alcohol and opium. His idea was first put into practice fifteen years after his death.

At the Royal Institution in London he installed the largest electric battery in the world, and experimented with electrolysis, the use of current to create a chemical action. By passing a current through various mineral solutions Davy became the first man to isolate many elements previously only found in compounds – potassium and sodium (previously thought to be the same element), calcium, magnesium, strontium, barium and boron. He also discovered that chlorine was an element, not a compound, and named it and its periodic companion iodine. Davy's interests were wide and his research methods, explosions apart, scientific and thorough. Had he completed his proposed magnum opus *Elements of Chemical Philosophy*, one contemporary commented, "he would have advanced the science of chemistry a full century".

# John Dalton

## (1766–1844)

## Atomic theory

John Dalton was a modern scientist with an old-fashioned mind. His contributions were sometimes compromised by outdated theories, wrong assumptions and inferior equipment, but his pursuit of "the ultimate particle" fully justifies him being called the Father of Modern Chemistry.

John Dalton was born on the northern fringe of the English Lake District, a Quaker stronghold. He was fortunate to have teachers who recognized his early scientific promise. His nonconformist religion, however, prevented him from attending an English university, where at the time only Anglican Christians were permitted to graduate.

His family's poverty and Quaker modesty taught Dalton a lifetime of frugality. It bred in him a self-reliance bordering on isolation, and may have rendered him less open to the new ideas emerging around him. In one of his books he wrote, "I have been so often misled by taking for granted the results of others that I am determined to write as little as possible but what I can attest by my own experience." For example he devised his own system of dots, lines and circles to represent chemical compounds and stuck to it even after Swedish chemist Jöns Jacob Berzelius had created the far simpler annotation of letters and numbers – for example $H_2O$ for water – which was eventually adopted internationally.

Dalton's great contribution to chemistry was the idea that atoms combine to make molecules and compounds in whole-number multiples: atomic theory. The existence of atoms had been proposed at several times in science history. The word itself is from the Greek for "indivisible" and the Greek philosopher Epicure was among the first to suggest that an object's properties came from the atoms of which it was composed. Although Epicurean philosophy was at odds with Christianity, it was a French priest, Pierre Gassendi, who revived the idea in the seventeenth century.

John Dalton entered the field through his studies of gases and his readings of Antoine Lavoisier's Law of the Conservation of Mass (that the products of a chemical reaction will have the same weight as the reagents); and Joseph Proust's Law of Definite Proportions (that the constituent elements of a compound will always be in the same proportion regardless of quantity).

He was interested in situations where the same elements could form more than one compound – for example nitrogen and oxygen can form three different gases; nitrous oxide ($N_2O$), nitric oxide (NO) and nitrogen dioxide ($NO_2$). His experiments proved that the ratios of each element as it occurred in relation to the other were simple whole numbers. For example, for 140g of nitrogen, $N_2O$ contains 80g of oxygen; NO has 160g; and $NO_2$ 320g. Oxygen is present then in the ratio 1:2:4 across the three compounds. In the case of the two iron oxides FeO and $Fe_2O_3$, oxygen is present in a ratio of 2:3. This phenomenon is Dalton's Law of Multiple Proportions.

This constant relationship proved that there was indeed some basic unit involved in the combinations – the atom. And the measurement of mass proved that every element had its own unique atom, constant in weight but different from all other elemental atoms. Dalton even went so far as to define the atomic weights of seventeen common elements.

Dalton did not know that some elements occur naturally as molecules (oxygen for example occurs as $O_2$, two atoms of oxygen), and assumed that elements will tend to combine in the simplest ways (so he thought water was HO, not $H_2O$). As a result his calculations were often wrong. But his theory was completely correct and, despite the proposals of Epicure, Gassendi and their followers, Dalton's is the first genuinely scientific theory of the atom.

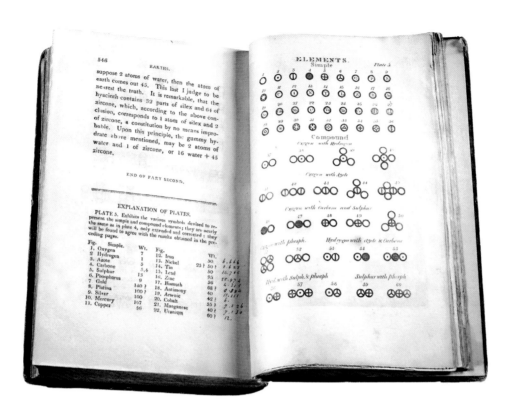

ABOVE: *An extract from Dalton's 1808 book,* A New System of Chemical Philosophy. *LEFT: A portrait of John Dalton by Thomas Philips from 1835. In 1793 he had taken up a post in Manchester, and his first paper to the Literary and Philosophical Society, in 1794, concerned his own visual problem, colour blindness.*

# Amedeo Avogadro

## (1776–1856)

## Avogadro's law of gases

Amedeo Avogadro clarified the difference between atoms and molecules, which had been a cause of confusion in the eighteenth century. His work with gases, however, went largely unnoticed until, in the years immediately before and after his death, the work of other scientists proved his hypothesis.

Avogadro's Law (as his hypothesis became known) states that under the same temperature and pressure, equal volumes of a gas will contain equal numbers of molecules, and therefore be of equal weight. In order to arrive at this idea, he had to be clear about the nature of molecules.

Some scientists still thought molecules were synonymous with atoms, the smallest components of atomic theory. But not all elements occur naturally as atoms; oxygen (O) for example only exists in the air as a pair of atoms combined as a molecule, $O_2$. And not all gases are single elements: carbon monoxide for example consists of molecules of one carbon (C) and one oxygen atom, CO; carbon dioxide molecules have two atoms of oxygen, $CO_2$.

Avogadro was the first man to distinguish between these three forms. Although he didn't use the word atom he coined the terms "molécule intégrante" (the molecule of a compound such as $CO_2$), "molécule constituante" (the molecule of an element such as $O_2$), and "molécule élémentaire" (an atom such as O or C).

Avogadro was building on Gay-Lussac's 1808 Law of Combining Gases and John Dalton's atomic theory, and his conclusions reconciled apparent differences between them. He argued that all gases were *"molécules intégrantes"* or *"molécules constituantes"*, not atoms. The extension of his Law meant that under the same conditions of pressure and temperature the relative molecular weights of any two gases are the same as their relative densities. This in turn made it possible to calculate the volume of gas in a container.

Avogadro made other contributions to chemistry. In the ten years following the publication of his hypothesis he correctly identified the molecular formula for water, nitric and nitrous oxides, ammonia, carbon monoxide, carbon dioxide, carbon disulphide, sulphur dioxide, hydrogen sulphide, hydrogen chloride, alcohol and ether.

Amedeo Avogadro is considered the founder of atomic-molecular theory, but it was almost half a century before he was recognized. He worked in relative isolation and was not part of the European chemistry community, which was dominated by the northern European states of Germany, France, Sweden and Britain. Eventually, however, work in the field of organic chemistry by Frenchmen Charles Frédéric Gerhardt and Auguste Laurent in the 1840s suggested that Avogadro might be right; and another Italian, Stanislao Cannizzaro, championed his Law and demonstrated its usefulness in chemistry four years after Avogadro's death.

Avogadro's Law greatly contributed to the understanding of gases. By the end of the century it led to the formulation of the kinetic theory of ideal gases, which explains the relationship between volume, pressure and temperature of gases and, among other things, Brownian Motion. In combination with the work of later scientists, including Michael Faraday, it led to the Avogadro constant, named in his honour, which is the total number of particles that make up a measure known as a mole. Moles may differ in mass and volume, but they all contain the same number of particles, be they ions, electrons, atoms or molecules. The number, the Avogadro constant, is $6.02214076 \times 1023$.

*OPPOSITE TOP: Avogadro achieved a rare feat by getting his law of gasses printed on an Italian postage stamp in 1956.*

*OPPOSITE BOTTOM: Amedeo Avogadro was born in Turin and spent his life there, teaching and researching at the University of Turin.*

# Hans Christian Ørsted

## (1777–1851)

## Electromagnetism

Driven by Emmanuel Kant's philosophy of unity in nature, philosopher-scientist Hans Christian Ørsted was determined to find a connection between electricity and magnetism. In doing so he kick-started the technological world we live in today.

Kant argued that science's job was to see and explain nature as a whole, a single entity. "Our diverse modes of knowledge," he wrote, "must not be permitted to be a mere rhapsody, but must form a system." His perspective was in a sense a throwback to earlier times when science was "natural philosophy", just one of many branches of learned thought. The two disciplines diverged over time. But in the final year of the eighteenth century, the Danish son of a pharmacist, Hans Christian Ørsted, was awarded his PhD for physics and aesthetics in the form of a dissertation on the works of Kant.

The following year Alessandro Volta invented the battery, which for the first time gave scientists a reliable and consistent flow of electricity with which to experiment. There was a rush to investigate the properties of this mysterious energy, and Ørsted soon devised a battery of his own. He was given public funds to travel through Europe visiting its research centres, and in Germany he met Johann Ritter, a fellow admirer of Kant and of Friedrich Schelling. Schelling was a philosopher who believed that scientists should concentrate on the search for "one absolute and necessary law" in which all natural phenomena were correlated and from which they could all be deduced.

Ritter's belief in a link between magnetism and electricity was infectious, although Ørsted rejected Schelling's argument that practical experimentation was less important than philosophy. On his return to the University of Copenhagen in 1806 he was appointed professor and set up experimental laboratories to conduct research into chemistry, physics, acoustics – and electrical currents.

At last, in 1820, he was able to demonstrate, by practical experiment, the evidence for the connection he had been seeking. During a lecture in April that year, he passed an electrical current through a wire, which caused the needle of a compass sitting next to it to deviate. The current was generating a magnetic field stronger than the Earth's, to which the compass responded. Ørsted had discovered electromagnetism. For this he was recognized by science's most prestigious award, the Copley Medal of the Royal Society in England.

Albert Einstein eventually confirmed that electricity and magnetism were inextricably linked aspects of the same phenomenon. Electromagnetism underpins subatomic physics, and electromagnetic radiation is behind such varied waveforms as X-rays, gamma rays, ultraviolet and visible light, microwaves, and radio and television broadcasts. Ørsted's discovery has shaped the modern world.

Hans Christian Ørsted made other contributions to science, in the realm of chemistry. He discovered piperine, the compound that gives black pepper its taste. And he was the first man to isolate pure aluminium. He was less successful, however, at distilling andronia and thelyke, two mythical substances which Hungarian chemist Jakob Winterl believed were the fundamental components of everything. The idea appealed to Ørsted's Kantian principles. If they were behind heat, light, electricity, magnetism, acids and alkalis, he reasoned, "we have the unity of all forces … and the former physical sciences thus combine into one united physics." Alas, they were not.

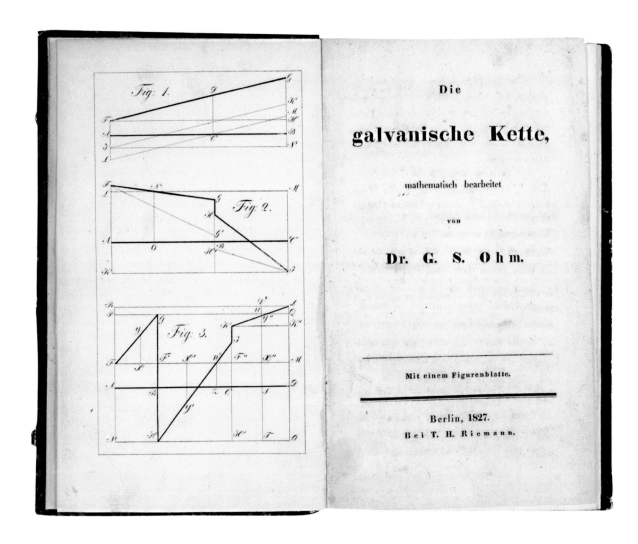

Die

# galvanische Kette,

mathematisch bearbeitet

von

# Dr. G. S. Ohm.

**Mit einem Figurenblatte.**

Berlin, 1827.
Bei T. H. Riemann.

*ABOVE: Ohm published* Die Galvanische Kette, Mathematisch Bearbeitet *in 1827. Despite the status that a published work brings, Ohm's college, the Jesuit Gymnasium of Cologne, did not appreciate his work and Ohm resigned from his position.*
*OPPOSITE: Georg Ohm was taught mathematics, physics, chemistry and philosophy by his father, who worked as a locksmith.*

# Georg Ohm

## (1789–1854)

## Ohm's Law

Student dropouts should take comfort from Georg Ohm, discoverer of Ohm's Law, one of the best-known rules in electrical physics. He spent more time in billiard halls and pubs than at lectures when he first went to university and quit altogether after only three terms.

Despite receiving a solid mathematical and scientific education at home from his father, Georg Ohm – like many students before and since – was easily distracted from his studies at Erlangen University in Bavaria. When his father found out that he was spending more time at the billiard halls and ice rinks of the town than at mathematics lectures, the young Georg was sent off to Switzerland, where he began teaching children mathematics instead.

This undemanding occupation gave him enough free time to study the works of the great mathematicians of the age – Euler, Lacroix and Laplace – as his Erlangen tutor had recommended. After three years he was ready to resume his place at university and study for a doctorate, which he received in 1811. He joined the staff as a lecturer but once again left quickly when his salary did not cover the costs of his rather hedonistic lifestyle. Dance halls were altogether more fun than lecture theatres, but also more expensive.

Instead he returned to school-teaching and eventually took up a post at the Jesuit school in Köln, where he was expected to teach physics as well as maths. Keeping one step ahead of his students in the school's well-equipped science department, Ohm's intellectual curiosity was at last engaged. Alessandro Volta's invention of the electrical cell in 1799 had precipitated a rush of scientific research into the properties and effects of electricity, and Ohm began to conduct his own experiments into it.

Ohm was interested in the conductivity of different metals. One of his early encounters with the notion

of electrical resistance was his observation that the electromagnetic field from a conducting wire decreased in strength as the length of the wire increased.

He was a prolific scientific author, and published his most important work, *Die Galvanische Kette, Mathematisch Bearbeitet* ("The Galvanic Circuit Investigated Mathematically"), in 1827. This contained his discovery through observation, that the current flowing through a conductor is directly proportional to the potential difference (voltage) and inversely proportional to the resistance.

The school in Köln was, however, unimpressed with Ohm's breakthrough, and in despair he resigned from his post to teach at the polytechnic college in Nürnberg. During his time there the scientific community at large woke up to his ideas, and in 1841 he was awarded science's most prestigious honour, the Royal Society's Copley Medal. Two years before his death Ohm, recognized at last, took up a position as professor of experimental physics at the University of Munich.

A committee of the British Association for the Advancement of Science met in 1861 to discuss the need for standard units of electrical measurement and chose to name them after the field's pioneers. They settled on the Volt (V) for potential difference, and the Ohma for resistance, soon shortened to Ohm ($\Omega$). The Ampere or Amp (A) was adopted for current at another convention in 1881. Thus Ohm's Law can be succinctly and mathematically expressed as:

$$A = \frac{V}{\Omega}$$

# Robert Brown

## (1773–1858)

## Brownian Motion

A widely respected botanist who made important contributions to his own science, Robert Brown was a devotee of the microscope and all that it could reveal about plants. Through the lenses of that instrument he also discovered the phenomenon known today as Brownian Motion.

"He is a Scotchman, fit to pursue an object with constance and cold mind." Thus was Robert Brown recommended for a scientific research voyage to Australia on board the HMS *Investigator*. He was born in Montrose on Scotland's chilly east coast and trained in medicine at Edinburgh University. He was far more interested in botany and when he should have been studying anatomy he was foraging for specimens in the Scottish Highlands.

He abandoned his medical course and when he was offered the position of naturalist on board the *Investigator* in 1800 he promptly accepted it. The *Investigator*'s mission was to discover whether Australia was a single island or an archipelago; and Brown's job was to amass, with his two assistants, as many specimens as possible of the flora of this new continent. He was assiduous in his task, collecting and recording around 3,400 species, most of them previously unknown to Europeans.

His subsequent publication on the plants of Australia and New Zealand won him a seat at the top table of botanists, first as personal librarian to the great English naturalist Sir Joseph Banks. When Banks's collections were donated to the British Museum in 1827, Brown went with them as the Museum's first ever Keeper of Botany.

In the course of his duties that year he examined a new plant brought back from North America only a year earlier by the renowned plant collector and fellow Scot David Douglas, best known for giving his name to the Douglas Fir tree. Brown studied its pollen in a suspension of water through his microscope.

What he saw were tiny particles, thrown out from the grains of pollen and dancing about like a crew of drunken sailors with the jitters. They were, we now know, organelles; Brown thought at first that they must be some form of life. But when he observed the same motion in inorganic particles he was forced to abandon that idea. He recorded his observations but made no further attempt to explain the phenomenon.

Nearly eighty years later, in 1905, it fell to a youngish Albert Einstein to come up with a convincing model for what Brown had seen. The particles were not moving, but *being moved* by the kinetic energy of individual water molecules which struck them at random. Brownian Motion, as the phenomenon has become known, occurs in closed systems in which the fluid (liquid or gas) is in equilibrium. Even in equilibrium – at rest – the molecules of fluids are in a constant state of motion. If a fluid contains other particles, for example Brown's organelles or the ash in smoke, then those particles will be buffeted by the moving but invisible molecules. The particles will appear to be moving randomly of their own accord, and that movement is Brownian Motion.

Brownian Motion was, as Einstein explained, visual evidence for the existence of atoms and molecules. But two thousand years earlier, the first person to express such a theory may have been the Roman philosopher Lucretius. In around 60 BCE he wrote:

"You will see [in sunbeams] a multitude of tiny particles mingling in a multitude of ways … their dancing is an actual indication of underlying movements of matter that are hidden from our sight … those bodies are in motion that we see in sunbeams, moved by blows that remain invisible."

*ABOVE RIGHT: The British Museum, where Brown was the first ever Keeper of Botany.*
*RIGHT: Robert Brown and his scientific champion, Albert Einstein.*

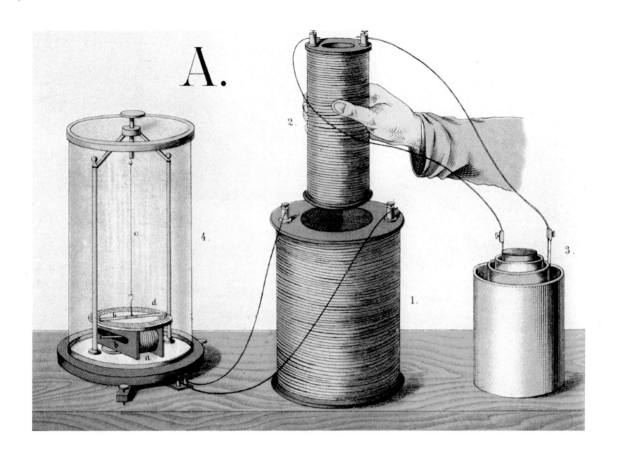

A.

ABOVE: *Faraday's electromagnetic induction experiment of 1882. The inner coil is connected to a liquid battery, the outer coil to a galvanometer.*
LEFT: *An 1844 daguerreotype photograph of Michael Faraday by the celebrated American photographer Mathew Brady.*

# Michael Faraday

## (1791–1867)

## Electromagnetic induction

Inspired by Humphry Davy, the finest chemist of the age, Michael Faraday turned from bookbinding to science. He was, some say, Davy's greatest discovery. Faraday's relentless pursuit of electricity took its toll on his health but gave the world the dynamo and the electric motor.

Michael Faraday, nineteenth-century beacon of electromagnetic science, entered the world of work at the age of thirteen, delivering newspapers for a local bookshop. The shopkeeper eventually took him on as an apprentice bookbinder and Faraday, largely self-educated, began to read. He read everything and in later life he would recall one book in particular – *Conversations on Chemistry* by Jane Marcet.

Marcet, the wife of a chemist, was a remarkable woman who wrote a long series of "conversations" on social and natural sciences, in the form of exchanges between two young women and their teacher. In a man's world she was rare but not unique – Mary Somerville and Harriet Martineau were also science writers, contemporaries and friends of Marcet.

Faraday was enthused by Marcet's book and when he had the opportunity to hear Humphry Davy give a series of lectures, he took it. His notes on them ran to 300 pages, which he bound into a book and sent to Mr Davy along with a request for a job. Davy was impressed by Faraday and found him a position as chemical assistant at the Royal Institution, preparing experiments for him.

During Faraday's bookbinding apprenticeship Davy had made his name by using electricity to isolate several elements. Now Faraday was invited to accompany the Davy family on a tour of the scientific hubs of Europe. It must have been a dream opportunity for the future pioneer. Back in London he used his precious spare time to conduct his own experiments into electromagnetic rotation, the phenomenon behind the electric motor.

In 1826 he instigated the Royal Institution's Friday evening lectures with the aim of popularizing science for the public, repaying the debt he owed to Jane Marcet's *Conversations* and Davy's lectures. Faraday himself proved to be one of the most gifted communicators of science at these gatherings.

Faraday published his findings in the matter of electromagnetic induction in 1831: the discovery that a current passed through one electrical coil could induce a current in a neighbouring coil. The first coil had effectively become an electromagnet. He repeated the experiment using a permanent magnet, and found that induction was possible by moving the magnet back and forth through a coil. This is the principle on which a dynamo operates.

By his work on induction Faraday transformed electricity from a scientific novelty act into a phenomenon with enormous practical potential. He spent the rest of his life exploring its properties, making particular contributions on the role of electricity on chemical bonds and the effect of magnetism on light. His experimental and theoretical legacy was the foundation on which James Clerk Maxwell constructed electromagnetic field theory.

The farad, a unit of electrical capacitance, was named after him. But Michael Faraday was a resolutely modest man. He twice turned down the offer of the presidency of the Royal Society and refused the honour of a knighthood, determined (he said) "to remain plain Mr. Faraday to the end".

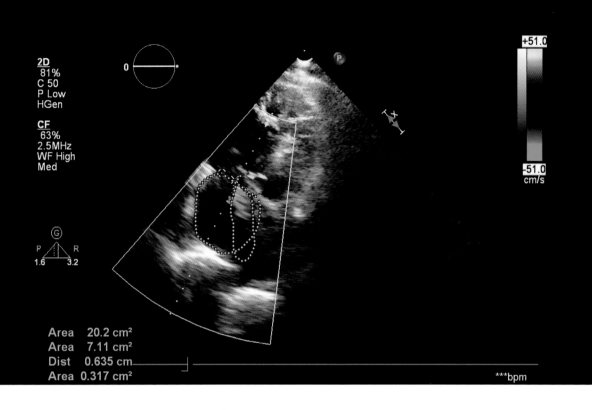

2D
81%
C 50
P Low
HGen

CF
63%
2.5MHz
WF High
Med

0

P    R
1.6    3.2

+51.0

-51.0
cm/s

Area    20.2 cm²
Area    7.11 cm²
Dist    0.635 cm
Area    0.317 cm²

***bpm

*ABOVE: An ultrasound image of the heart using the Doppler mode.*
*LEFT: The Zillertalbahn mountain railway in the Austrian Tyrol.*
*OPPOSITE: Christian Doppler was born in Salzburg, Austria, in a house adjacent to a former residence of the Mozart family.*

# Christian Doppler

## (1803–1853)

## The Doppler Effect

Christian Doppler discovered the Doppler Effect not in relation to the sound of passing express trains, or the approach of a nineteenth-century fire tender, but to the changing colours of binary stars.

Doppler, a citizen of the Austro-Hungarian Empire, made his name with a lecture to the Royal Bohemian Society of the Sciences in 1842. He subsequently published it as *Über das farbige Licht der Doppelsterne und einiger anderer Gestirne des Himmels* ("On the coloured light of the binary stars and some other stars of the heavens").

The Doppler Effect is easier to hear or see than it is to explain. Light of a certain colour travels as waves of a certain frequency, with the crests of the waves equally spaced apart. If the source of the light is moving, then the waves no longer reach your eye at the same frequency. The position of the crests changes in relation to each other and the light may appear as a different colour or frequency.

Sound travels in waves too. The engine of an express train makes a constant noise, but as the engine approaches, newer sound waves from it have less distance to travel before they strike your ears. As a consequence, the sound waves reach you closer together and, at this higher frequency, the engine sounds higher pitched than it is. As it passes you and starts to get further away again, the frequency of the waves reduces and the tone of the engine decreases to below its actual sound. If a train were travelling in a circle with you at the centre, there would be no Doppler Effect because all the sound waves would be crossing the same distance to reach your hearing.

It was a train which first demonstrated the audio Doppler Effect, three years after Doppler's publication

which focused on light waves. Noted Dutch chemist Buys Ballot employed a small brass band to travel on the Utrecht to Amsterdam railway line and play a single constant note. He measured its pitch as the train approached, passed and receded from him and confirmed that Doppler's theory worked as well for waves of sound as of light.

For most of us the Doppler Effect is just a curiosity associated with the passing of fire engine sirens. For science, however, Doppler's explanation is of profound interest in many fields of study. It has been shown to apply to electromagnetic waves too, and in astronomy the phenomena of red shift and blue shift are explained by rises and falls respectively of electromagnetic frequencies.

The effect can also be exploited in the use of radio waves, for example in the control of mobile robots, or in communication with fast-moving satellites. Radar waves fired from police speed guns take the Doppler Effect into account when calculating the speed of a receding vehicle.

The Doppler Effect has applications in medical science too. An ultrasound scan of the heart, an echocardiogram, makes use of it to determine the direction and volume of blood flow, a useful tool for early diagnosis of cardiovascular problems.

# James Prescott Joule

## (1818–1889)

## Laws on conservation of energy

James Joule overturned the belief, held since Antoine Lavoisier proposed it in the 1770s, that heat could neither be created nor destroyed. It was an uphill battle for an outsider to convince the scientific community: Joule's enquiries into the subject were inspired by his love of beer.

To be precise, James Joule was a brewer who took over from his father in running the family brewery in north-west England. He had shown a fascination with electricity since his youth, when he and his brother would administer electric shocks to the family servants to observe the effects. At the brewery, he introduced electric motors to replace the steam engines that had previously powered it, and he wanted to know which power source was more efficient for his business.

In purely financial terms it proved to be the steam engines, because the zinc in electric batteries (in the days before mains electricity) was far more expensive than the coal in a boiler. But in the process of discovering this Joule devised increasingly accurate ways of measuring the work done by different forms of energy. The gauge was what it took to raise a one-pound (454g) weight to a height of one foot (30.5cm). He called this the foot-pound.

Joule's research led him to wonder about the nature and measurability of work. He noted that heat and energy often occurred together, whether in the temperature of an electrical conductor or in water being churned by a paddle. Joule came to believe that heat, light, electricity and mechanical effort are all just different forms of the same energy and that one can be converted into another. This is the basis of the First Law of the Conservation of Energy.

It directly contradicted Antoine Lavoisier's theory of heat, which held that heat was a property of a substance called caloric, a "subtle fluid" of which there was a fixed amount in the world. Caloric could be neither destroyed nor created but simply passed from one material to another. First expressed in the 1780s, Lavoisier's theory was the orthodoxy in the 1840s.

The scientific establishment was therefore sceptical. Joule found it difficult to get published and his lectures were often heard only in polite silence. What did a brewer know of scientific theory and practice? In fact his appreciation of the temperature-sensitive art of brewing gave Joule an added advantage in his experiments.

At one such lecture two electrical heavyweights, Michael Faraday and William Thomson, were in the audience. Both were impressed with Joule's ideas and the respect of Thomson, the future Lord Kelvin, went a long way to making them acceptable. Joule and Thomson worked together on the measure of temperature that became known as the Kelvin scale, and they observed the Joule-Thomson Effect in which a gas under pressure rises in temperature.

Caloric Theory persisted in school textbooks for most of the century; but Joule eventually convinced his peers. He was awarded the Royal Medal (in 1851) and the Copley Medal (in 1870) by London's Royal Society. Joule lived long enough to see his name adopted as the unit of measurement of energy, a variation of his own foot-pound. One joule is the work required to produce one watt of power for one second.

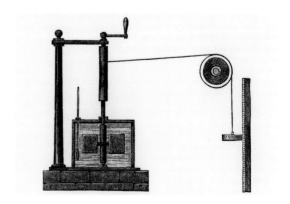

LEFT: *Joule was taught mathematics by atomic theorist John Dalton.*
OPPOSITE: *Joule's apparatus for studying heat conservation and loss.*
BELOW: *The family brewery with the trademark red cross.*

ABOVE: *Oymyakon in Siberia is the coldest permanently inhabited place on Earth and is found in the Arctic Circle's Northern Pole of Cold. In 1933, it recorded its lowest temperature of -67.7°C. Quite warm on the Kelvin Scale.*

LEFT: *William Thomson became a professor at Glasgow University at the age of 22.*

# William Thomson, Lord Kelvin

## (1824–1907)

## Definition of absolute zero

The idea that there is an absolute zero, the lowest possible temperature, has been around since at least the seventeenth century, when natural philosophers argued among themselves about which of the four elements – earth, air or water (not fire of course) – was capable of reaching it.

French physicist Guillaume Amontons devised a temperature scale in 1702 by which water boiled at 73° and froze at 51.5°, which would have placed his 0° at around -240° Celsius. In Poland, Daniel Gabriel Fahrenheit made 0° the temperature at which a brine of water and ammonium chloride froze. The other fixed point of his scale, devised in 1724, was the temperature of the human body, which he set at 96°.

In 1742 Anders Celsius proposed a more rational scale in which a tidy 100° separated the two events, giving a logical 0° and 100° as his fixed points. Boiling and freezing water were at least observable commonplace events. The reason none of these great men set their 0° at absolute zero is that none of them knew exactly where it was. The atomic chemist John Dalton, for example, thought it might be around -3000°C, while his contemporary Pierre-Simon Laplace thought it could be as high as -1500°C.

Because thermometers relied on the expansion and contraction of either mercury or air, their accuracy was limited to the responses of those substances to changes in temperature. Their purity, or the use of other materials altogether, would result in different movements along the scale. The only truly fixed point, then, was absolute zero, wherever that was.

William Thomson, the Irish-born professor of natural history at Glasgow University, was determined to find a scale that was independent of any materials used in measuring temperature. Instead of measuring existing extremes of temperature therefore, he approached it with the theoretical logic of a scientist – three scientists

in fact. He corresponded with James Joule about the mechanical equivalent of heat, and with Henri Victor Regnault on the thermal properties of gases; and he studied Nicolas Carnot's theory of motive power from heat.

Thomson observed that for a given volume of any gas, the relationship between temperature and pressure remains constant. Plotting one against the other on a graph produced a straight line. He proposed that when the pressure reached zero, it would be because there was no heat to generate energy, and therefore the temperature at that point on the graph would be absolute zero.

He thus determined a value for absolute zero of -273°C. With the benefit of greater experimental accuracy we now know that 0°K, the absolute zero of the Kelvin scale, is -273.15°C, or -459.68°F. The Kelvin Scale advances in degrees Celsius. Another scale, which also starts at absolute zero but advances in degrees Fahrenheit, is called the Rankine scale, after another Glasgow University physicist, Macquorn Rankine.

Setting a value for absolute zero spurred the experimental scientific community on to try to reach it. The average temperature of the universe is 2.73°K, or -270.42°C (-454.756°F), but a temperature of approximately 1°K was observed in the Boomerang Nebula; and scientists have artificially reached 0.0000000001°K by slowing the nuclear activity within a sample of the element rhodium. It is currently accepted that engineering a temperature of absolute zero is impossible.

# Rudolf Virchow

## (1821–1902)

## Cellular pathology

Even after the discovery that all animal tissue was composed of cells, men of medicine clung to irrational theories of illness and disorder with their roots in the "golden age" of classical Greece and Rome. Rudolf Virchow's rational approach to disease ushered in the modern medical era.

Medical science was slow to shrug off the philosophical approach of Galen, the ancient Roman who made the earliest written studies of the human body. He subscribed to the view that the body consisted of four "humours" – blood, phlegm, and yellow and black bile – which must remain in balance for good health.

A person's character was thought to be derived from their dominant humour. Even today we use words derived from that school of thought. "Sanguine" is derived from the Latin for "blood". Someone with too much phlegm is described as "phlegmatic".

Such was the pervasive acceptance of Galen's primitive understanding of the human body that even after Theodor Schwann established the existence of cells, he believed that they were created by a mythical bodily fluid called blastema. If the body was somehow out of balance, the blastema would produce diseased cells.

We now use the term blastema to describe a set of cells which allow certain animals to regenerate organs and limbs, thanks to Rudolf Virchow's maxim that "all cells come from cells". This countered the commonly held belief that lower forms of life could come spontaneously from non-living matter – for example, that maggots simply appeared on rotting meat. Virchow was echoing the Italian biologist Francesco Redi, who argued that "every living thing comes from a living thing".

Virchow was a German physician who was drawn to anatomy, pathology and the microscopic study of cells during his training in Berlin. At the city's Charité

teaching hospital he learned to conduct detailed autopsies and was the first man to identify, describe and name leukaemia (the name means "white blood" in Greek), a disease of under-developed blood cells.

From this he developed the theory that diseases come from changes in previously healthy cells and that different diseases affect different groups of cells. He advised doctors to examine diseased cells in order to diagnose accurately. At the time it was still the practice to consider only a patient's symptoms, which was scarcely more reliable than diagnosing an excess of a particular Galenian humour based on the pallor of the patient's skin.

Virchow went on to describe and name many diseases for the first time, based on their cellular pathology, including embolism, thrombosis, chordoma, and ochronosis. His approach met with considerable resistance from the medical establishment. Faced with refusals by journals to publish his work, he neatly sidestepped them by launching his own pathology magazine, the *Archive for Pathological Anatomy and Physiology and Clinical Medicine*. Defiantly modern in its insistence on rigorous anatomical research, the journal still exists, now published as *Virchow's Archives*.

Virchow was also an active politician who campaigned for improved public health after he was shocked by the impoverished conditions in an area hit by a typhus outbreak. "Medicine is a social science," he said, "and politics is nothing else but medicine on a large scale."

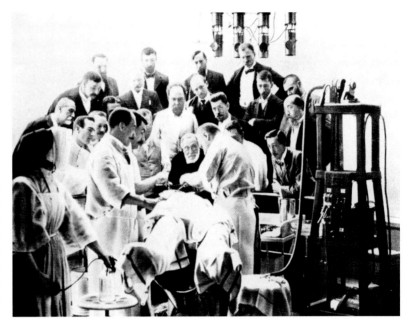

*RIGHT: German pathologist Rudolf Virchow (middle, in a dark suit) supervises cranial surgery in a Paris hospital.*
*BELOW: Computer illustration of an abnormal white blood cell from a patient with hairy cell leukaemia, a rare form of the disease.*

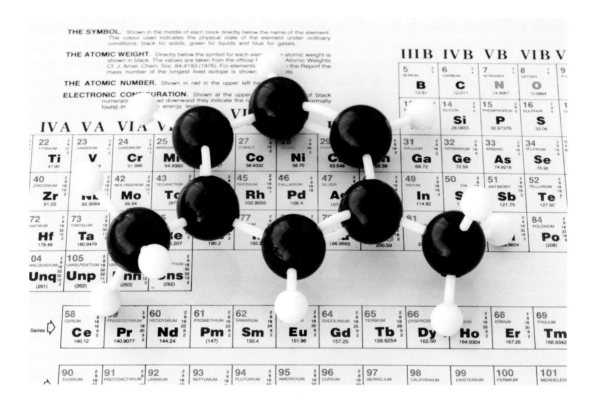

TOP: *A plastic model of a 1,3 dimethylbenzene molecule, one of the xylene isomeres.*

ABOVE: *An East German postage stamp honouring German chemist Friedrich August Kekulé.*

OPPOSITE: *Archibald Scott Couper came to the same conclusion as Kekulé but his supervisor failed to deliver his paper in time.*

# August Kekulé & Archibald Scott Couper

## (1829–1896 / 1831–1892)

## Atom bonding and carbon rings

This is a story of luck and dreams. Two men, unaware of each other's research, made the same discovery at the same time. But in the world of scientific priority there can only be one winner. One man went on to fame and fortune; the other retired disappointed, angry and broken.

August Kekulé was an organic chemist from Darmstadt in Germany. At university he switched from architecture to chemistry after hearing the lectures of charismatic scientist Justus von Liebig in 1847. In travels around the chemistry centres of Europe he was profoundly influenced by Alexander Williamson, a structural chemist in London.

Structural chemistry deals with the manner in which atoms bond together to form molecules. In the mid-nineteenth century, scientists were making great strides in discovering the composition of molecules – which elements were present and in what quantities. But the precise mechanism by which they bonded was still a matter of speculation.

A new theory was emerging, the concept of atomic valence, according to which the atoms of each element had a fixed number of bonds available for connection to other atoms. Oxygen, for example, has a valence of 2, having two such bonds; hydrogen has one. So a molecule of water, $H_2O$, has a structure H-O-H in which oxygen's two bonds each pair with the single bond of one of the hydrogen atoms. It's easy enough to work out the structure of such relatively simple molecules, but complex organic compounds – hydrocarbons for example – present many possible variations.

Kekulé subscribed to the atomic valence theory advocated by Williamson, Charles-Adolphe Wurtz and others. He made an early breakthrough in 1857 when he

determined that carbon had a valence of 4; and in May the following year he published a paper outlining his discovery that carbon atoms could use some of their bonds to combine with each other in circular structures and still have enough bonds left to combine with other atoms. This and their relatively high valence explained why compounds could contain many carbon atoms and exhibit such a variety of combinations with hydrogen and oxygen. It was a huge step forwards in structural chemistry, and paved the way for Kekulé's later discovery of the structure of benzene, $C_6H_6$.

Meanwhile a Scottish chemist, Archibald Scott Couper, was working in the Parisian laboratories of one of the great evangelists for atomic valence, Charles-Adolphe Wurtz. Couper came independently to the same conclusion about carbon rings as Kekulé. He asked Wurtz to present his paper on the subject to the French Academy; but whether through procrastination or misunderstanding, Couper's paper was only read in June 1858, a few vital weeks after Kekulé's was published.

Couper, disappointed and angry, yelled at Wurtz and was expelled from the latter's laboratory. He retired to his native Scotland where he suffered a series of mental health problems and lapsed into inactive depression, cared for in the last thirty years of his life by his mother. To Kekulé went the credit, and the many honours which followed, including ennoblement by Kaiser Wilhelm II and the gratitude of chemists ever since.

# Charles Darwin

## (1809–1882)

## Theory of evolution by natural selection

The fact that man and other animals have evolved over millions of years is one of the most important and best-known discoveries in science. It was such a radical idea that Charles Darwin delayed the publication of his theory for more than twenty years after it first occurred to him.

The dominant discourse concerning life on Earth in the early nineteenth century was that God had established a natural order of creatures, some superior to others. Any variations in species had been ordained by Him when He created the world. A young Charles Darwin had heard of emerging theories of evolution based on studies of fossils but took little interest in them.

He was, however, interested in the natural world. While studying medicine at Edinburgh University he spent more time investigating marine invertebrates than attending surgery demonstrations, which he found distressing. Despite transferring to Cambridge University he continued to neglect his studies, preferring instead to collect beetles and plants. A botany professor recommended him for a position as naturalist of the survey ship HMS *Beagle*, and he set sail two days after Christmas, 1831.

The intended two-year voyage ultimately lasted five years, during which Darwin collected fossils as well as specimens of plant and animal life. On his return he asked the scientific institutions of London to identify the items he had brought back. They confirmed that, among other things, the twelve finches he had brought back were not just varieties of a species but each one a distinct species. The possibility began to occur to him that, as he wrote in a notebook at the time, "one species does change into another".

This was survival of the fittest, the adaptation of the most suitable for any given environment. It implied that the natural order was not fixed, by God or anything else. It could change, and indeed had repeatedly changed, in response to circumstances.

Darwin knew that he must be absolutely certain of his theory, and proceeded cautiously, building his reputation as a biologist while he conducted further research. The potential hostility which his views might provoke was confirmed when an anonymous 1844 publication, *Vestiges of the Natural History of Creation*, contained a crude version of evolutionary theory and provoked the fury of the clergy.

Time was not on his side, and other scientists began to converge on his ideas. One of them, Alfred Russel Wallace, delivered a paper in 1855 which coincided to a large extent with Darwin's own thinking. It persuaded Darwin that it was time to get into print, and *On the Origin of Species* finally appeared in 1859, more than twenty years after his voyage on the *Beagle*.

To avoid controversy Darwin's text made no direct point about the evolution of his own species, *Homo sapiens*. All the same he faced the outrage of those convinced of the traditional view, and the merciless lampooning of those who found it simply laughable that mankind could be descended from the apes.

The scientific community, already open to the idea, was quick to embrace the idea of evolution but slow to credit Darwin's theory of natural selection; that only achieved mainstream acceptance in the 1930s. It now forms the cornerstone of the life sciences, although it has found new sceptics since the recent resurgence of fundamentalist Christianity.

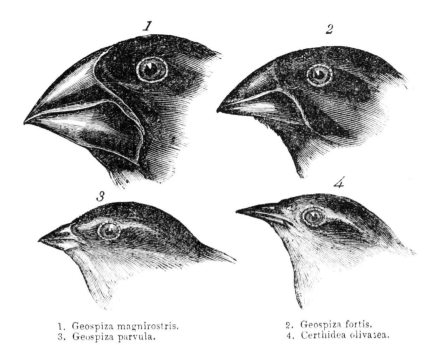

1. Geospiza magnirostris.
3. Geospiza parvula.

2. Geospiza fortis.
4. Certhidea olivasea.

ABOVE: *An illustration of four of the twelve Galápagos finches that Darwin observed on his voyage.*
LEFT: *A young Charles Darwin as he looked before his five-year expedition on HMS Beagle.*
OPPOSITE: *A conservative British press was keen to lampoon Darwin's ideas with cartoons of the great scientist as a monkey.*

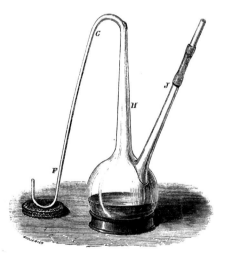

TOP: *Exhibits from the Pasteur Museum in the Institute Pasteur, Paris.*
ABOVE LEFT: *Louis Pasteur discreetly asked his family not to reveal his notebooks after his death.*
ABOVE RIGHT: *Using a swan-necked flask, Louis Pasteur demonstrated that microbial fermentation is caused by airborne organisms. It was only when the neck of the flask was broken that a change in the solution occurred.*

# Louis Pasteur

## (1822–1895)

## Germ theory

In the early nineteenth century, many people, including doctors, still believed that disease was caused by miasmas, pockets of poisonous air containing rotting particles. Disease was spread, it was thought, not from person to person but by being in an area on which a miasma had settled.

There had been periodic suggestions through the ages that disease was somehow germinated by malignant pollen in the air. In the second century, Greek physician Galen speculated that plagues were spread by "certain seeds". This idea resurfaced in the ideas of sixteenth-century Italian Girolamo Fracastoro, who wrote of *seminaria morbi*, or seed-based diseases.

Roman statesman Marcus Terentius Varro unwittingly hit the nail on the head, although his speculations were ignored for 1,600 years. He suggested in 36 BCE that swamps contained "certain minute creatures which cannot be seen by the eyes, which float in the air and enter the body through the mouth and nose and there cause serious diseases".

Not until the invention of the microscope was it possible to confirm his theory. Athanasius Kircher, a Jesuit priest in seventeenth-century Germany and one of the first microbiologists, examined rotting produce and the blood of plague victims, where he saw otherwise invisible "worms" which he believed could be the cause of infection. He was right, although scientists now think he may only have seen blood cells at the time.

Subsequent experiments by others began to make progress against the prevailing miasma theory. In Vienna, Ignaz Semmelweis noticed that the death rate in his maternity ward was higher when doctors who had come straight from autopsies were involved in the delivery. By getting them to wash their hands he reduced the rate from 18% to 2% in a year. In London, John Snow traced the source of an epidemic to a contaminated water pump. By simply removing the handle, he forced people to get their water from elsewhere and ended the outbreak.

Although such successes pointed to the germ theory of disease, it wasn't until Louis Pasteur's high-profile results against microbe-borne disease that the miasmas' days were numbered. Pasteur was a professor of chemistry whose life-saving research began when he investigated the fermentation process of different yeasts. To retard unwelcome fermentation he rapidly heated the liquid to around 60°C and slowly cooled it. The process became known as pasteurisation and is still widely applied to beer and milk.

That work led him to examine diseases in grapes and in silkworms, which were a threat to those two economically vital French industries. In the case of silkworms, the cause was hereditary micro-organisms and Pasteur solved the problem by destroying the eggs of moths known to carry them. He ended the epidemic of wine diseases by pasteurising the wine, killing off the micro-organisms in the grapes.

Pasteur next worked on the incidence of cholera in hens. After a mistake in the laboratory he discovered that hens which recovered after being treated with a weakened dose of the disease could no longer be infected even with a full-strength dose. Excited by this result he then succeeded in immunizing cattle against anthrax by the same technique. Finally he experimented with inoculating human beings against rabies. In his field test of 350 subjects, only one developed the disease.

Pasteur's work in microbiology was the proof that diseases were caused by germs. His painstaking development of the theory, accompanied by successful remedies, transformed medical practice and has saved millions of lives every year ever since.

# James Clerk Maxwell

## (1831–1879)

## Theory of electromagnetism

Electromagnetism lies behind much of the technology of the modern world. As nineteenth-century scientists struggled to understand the link between magnetism and electricity, one man had the capacity to see and define the whole field in a series of equations, laying down the foundations for the future.

James Clerk Maxwell showed early indications of his relentlessly inquiring mind. He had an insatiable childhood curiosity about how things worked, from electricity in the doorbell of the family home to the plumbing in the pantry. At the age of only fourteen his geometric paper on oval curves was presented to the Royal Society of Edinburgh.

At Edinburgh University in 1855 he experimented with the human perception of colour and demonstrated that white light could be made from red, green and blue light. In 1861 he developed the world's first colour photograph using a series of red, green and blue filters. At Cambridge University in 1857 he had correctly proposed that Saturn's rings consisted of particles, something which was only confirmed in the late twentieth century by NASA probes.

Maxwell's greatest work was done at King's College London where he took up a post as professor of natural philosophy in 1860. There he came into contact with an elderly Michael Faraday who had made early studies of the "lines of force" surrounding electricity and magnetism. Maxwell had already written about those lines, supporting Faraday's inference that the two forces were related.

Now he developed Faraday's ideas, experimenting with the properties of both and concluding that Faraday's "lines" were electromagnetic waves, travelling at the speed of light. That speed, he thought, was no mere coincidence but implied that light was a form of the same energy. So, as we now know, are other forms of electromagnetism such as X-rays, microwaves and radio waves; cornerstones of modern life.

Maxwell published his conclusions in a foundational work, *On Physical Lines of Force*, in 1861. It contained all that was now known about electromagnetism, and distilled it in twenty mathematical equations defining each of twenty variables. The differential equations not only summed up electromagnetic theory but predicted future discoveries in the field, for example in outer space.

Today, further distilled as four core equations known as Maxwell's Equations, they neatly sum up electromagnetism theory. Maxwell had a mind that was as strong on concept and visualization as it was on mathematics. When he died, too young, in 1879 – at the same age, 48, and of the same disease, abdominal cancer, as his mother – the world was robbed of one of the finest scientific minds the world has ever known. He had just begun to correspond with another, Willard Gibb, in the related field of thermodynamics; and who knows what advances those two might have made when working together?

*RIGHT: James Clerk Maxwell as a young man. Maxwell was voted the third greatest physicist of all time, behind only Newton and Einstein. BELOW: The rings of Saturn. Apart from his work on electromagnetic radiation, Maxwell also studied the motion of molecules in a gas, and showed that Saturn's rings must be composed of particles.*

*TOP: Rudolf Clausius's most famous statement on entropy was published in German in 1854: "Heat can never pass from a colder to a warmer body without some other change, connected therewith, occurring at the same time." ABOVE: The droplets of an aerosol moving from low entropy to high entropy.*

# Rudolf Clausius

## (1822–1888)

## Entropy

Thermodynamics – the study of heat, work, temperature and energy – began to take shape at the start of the nineteenth century. As scientists sought to get to grips with the new discipline, it fell to Rudolf Clausius, a Polish physicist, to refine its laws and identify the concept of entropy at its heart.

With the industrial revolution underway, there was pressure on industrial machinery to be as efficient as possible. French physicist Lazare Carnot published *Fundamental Principles of Equilibrium and Movement* in 1803, which prompted a scientific debate about energy that was generated by machines but not harnessed for work. Loose components, friction, misaligned gears, simple radiation of heat – they all dissipated energy intended to do the work at hand, and persuaded Carnot that the dream of perpetual motion was unattainable. Carnot's acknowledgement of this lost energy was a rough early statement of what became the Second Law of Thermodynamics.

Further study was made of lost work, not least by Carnot's son Sadi, who conceived of an engine whose loss of energy through heat could simply be reversed by reversing the machine. Sadi Carnot concluded, wisely, that the same heat would be lost all over again by such a process. Working within the orthodoxy of the day, Carnot believed that heat came in the form of a fluid substance called caloric which could move from one body to another. Although that theory was wrong, the loss of caloric was analogous to the future concept of entropy, and was a further illumination of the coming Second Law.

Rudolf Clausius discussed Carnot's ideas in an 1850 publication, *Über die bewegende Kraft der Wärme …* ("On the Moving Force of Heat and the Laws of Heat which may be Deduced Therefrom"). He found some conflict between the loss of heat as accepted by Carnot and the principle of conservation of energy. In trying to resolve the contradiction he reconsidered the rules of thermodynamics and the reality behind them. Over the next fifteen years he restated them in different ways and tried to explain the principle behind heat loss. In 1865 he arrived at these two Laws:

• The energy of the universe is constant.
• The entropy of the universe tends to a maximum.

The word "entropy" was his own invention, loosely translated from the Greek for "transformation within". It describes the potential for energy loss within a closed system (one not open to outside influences and effects), referred to scientifically as disorder. Entropy is measured by the number of ways in which all the components of a system can be reordered into different relationships with each other.

A good example is a can of air freshener. As long as the particles of air freshener are packed together in the can (a closed system) there is little scope for rearrangement, and therefore low entropy; but when you press the button and spray the contents into the air they disperse in a disorderly fashion throughout the room, a much larger closed system with high entropy. In the universe, the largest possible closed system, entropy is naturally at its highest.

Entropy, Clausius's conceptual gift to the scientific world, can also be applied to information systems, describing the loss of some signal in a telephone line for example, or of data in IT systems. In a human being, maximum entropy, the complete loss of energy, results in death.

# Gregor Mendel

## (1822–1884)

## Genetics

Prone to depression throughout his life, Gregor Mendel found solitude and peace in the vegetable garden of his monastery. There he also found the secret of life, or at least the way in which characteristics are inherited or discarded by successive generations. It took geneticists thirty years to catch up with him.

Gregor Mendel was born in a part of the Austrian Empire that now forms the southeastern Czech Republic. His father hoped he would help with the running of the small family farm; and perhaps farming practice did affect Mendel's future path. Mankind has played with genetics for thousands of years by cross-breeding crops and livestock to improve their yield. But although as a boy Mendel enjoyed bee keeping and gardening, he chose instead to study sciences in the nearby university town of Olomouc.

Instead of returning to the farm after graduation, Mendel entered the monastery of St Thomas in another Czech university town, Brno. The move was in part a solution to his financial problems: it meant he could continue to study without having to worry about the cost of accommodation and food.

The monastery sent him to Znojmo on today's Czech-Austrian border to teach in a school, and then to Vienna to continue his science studies full time. There he learned mathematics and physics from Christian Doppler (of the Doppler Effect) and botany from Franz Unger, an early proponent of the microscope who was conducting research into a pre-Darwinian evolution theory.

From Vienna, Mendel returned to the monastery in Brno, where he was again assigned to teaching duties in a local school; but the abbot also gave him permission to plan and undertake a large-scale botanical experiment in the monastery garden.

Mankind had a limited understanding of genetics in Mendel's time. It was clear that traits were inheritable – facial features for example, and certain health conditions. But the process was considered to be an even blend of

the characteristics of the parents in which extremes were somehow averaged out. Mendel's new experiment proved that this was not the case.

Mendel began a programme of cross-fertilization between different species of edible pea. The plants were easy to breed and quick to grow, which would speed up his harvesting and analysis of the results. There are many pea varieties and he chose seven with distinct characteristics of size, colour, texture and flower position. He discovered the principle of dominant and recessive genes. For example, a cross between yellow and green pea plants produced only yellow peas in the next generation; but in the following generation of hybrids some green pea plants reappeared among the yellow, in the ratio of approximately 1:3. The yellow gene was dominant, the green one recessive.

Mendel proposed two genetic Laws – the Law of Segregation, proposing dominance and recession rather than a blending of genes; and the Law of Independent Assortment, which stated that traits are passed on individually, not *en masse*.

Although Mendel did present his work in a lecture in Brno, and published it in papers, he did not promote it. He had little time after his monastic duties, especially when he was appointed abbot of St Thomas's. His work languished in obscurity until, 35 years later, other geneticists began to reproduce his results, notably the Dutch botanist Hugo de Vries and the English biologist William Bateson, who coined the word "genetics" in 1905. Through his pioneering study Mendel posthumously prompted the discoveries of chromosomes and the structure of DNA.

PLATE IV.—MENDELIAN INHERITANCE OF THE COLOUR OF THE FLOWER IN THE CULINARY PEA

Two flowers of a plant
of a pink-flowered race.

Two flowers of a plant produced by
crossing the pink with the white.

Two flowers of a plant
of a white-flowered race.

*TOP: On the centenary of his death in 1884, the German Post Office issued a commemorative
stamp for Gregor Mendel.*

*ABOVE: "Mendelian inheritance of the colour of the flower in the culinary pea", from* Breeding and Mendelian
Discovery *by A. D. Darbishire (London, 1912).*

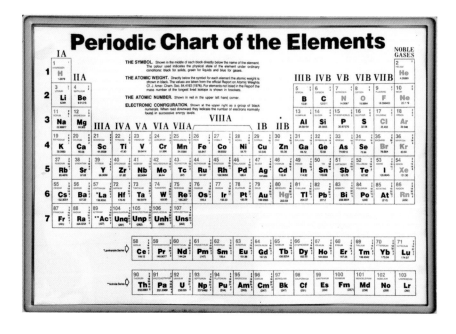

*TOP: Dmitri Mendeleev won the Davy Medal (for an outstanding discovery in any branch of chemistry) as early as 1882, but the Nobel committee ignored the Russian.*

*ABOVE: A chart that is replicated in school chemistry labs the world over.*

# Dmitri Mendeleev

## (1834–1907)

## The Periodic Table

The driving factor throughout the life of Dmitri Mendeleev was a desire to create order out of chaos. A compulsive organizer and list maker, Mendeleev discovered the Periodic Law and devised the ultimate list, the Periodic Table of Elements.

He was born in the ancient Siberian capital Tobolsk, where his father died when he was thirteen. His mother attempted to reopen the family's old glass factory, but it burned down a year later and she travelled west across Russia with her youngest, Dmitri, in search of better fortune. They settled in St Petersburg, a world away from Siberia, where Dmitri's late father had studied and where, in 1850, his mother died. Dmitri Mendeleev turned to science to make sense of his fractured world.

As a teacher at St Petersburg University he solved the problem of a lack of good chemistry textbooks in characteristic fashion, by writing them himself. While writing chapters on the properties of two groups of similar elements – the halogens and the alkali metals – Mendeleev noticed that the atomic weights of the elements within each group were close to each other.

At the time, most attempts to arrange the elements stemmed from the theory of William Prout, an English chemist, that they all developed from the same single source. Most approaches therefore looked more like family trees than lists or tables. Mendeleev, by contrast, believed that each element was unique and connected to others only in compounds.

His attention had been drawn to atomic weights in a paper presented by Stanislao Cannizzaro, an Italian chemist, at a conference a few years earlier. Now it struck him that listing the elements by weight, and therefore properties, might be a good way to organize them for teaching purposes. As he constructed this list, he formulated what he called the Periodic Law: as he put it, that "elements arranged according to the value of their atomic weights present a clear periodicity of properties". Periodicity means simply sharing similar properties in the same circumstances.

Not content with making his list, Mendeleev then set about presenting the information in diagram form, and came up with the Periodic Table of Elements, which we still use today. It is much larger today than when Mendeleev incorporated the seventy elements then known. But such was the logic of his table that he could use gaps in it to predict the discovery of new elements and their likely properties. When gallium was discovered in 1875, scandium in 1879, and germanium in 1886, the validity of his approach was indisputable.

Mendeleev was twice nominated for the Nobel Prize for chemistry, in 1905 and 1907. Both times he was denied the accolade by behind-the-scenes pressure from Svante Arrhenius, a Swedish scientist who had himself won the prize in 1903 and of whose theories Mendeleev had been critical.

Most of Mendeleev's scientific life was spent in the service of his country. He took a close interest in Russia's agricultural efficiency, and studied the oil and gas industries of America and Azerbaijan. Appropriately for such a consummate organizer, he was appointed director of Russia's Board of Weights and Measures in 1892. In that post he was responsible for introducing the metric system to the country; but he did not (as is sometimes claimed) set the 40° standard for vodka production.

# John Ambrose Fleming

## (1849–1945)

## Fleming's right-hand and left-hand rules

Ambrose Fleming turned the disadvantage of deafness into a positive force. His poor hearing deprived him of distractions and enabled him to focus on the scientific problems at hand. His inventive mind devised two very visual rules still used by electrical engineers today.

Fleming came up with his Right-Hand and Left-Hand rules as an aid for his students at University College London, where he was the institution's first ever professor of electrical engineering. As a student himself at Cambridge University, he had been among the last cohort to hear lectures by James Clerk Maxwell, pioneer of electromagnetic radiation, in 1877.

Fleming corresponded with Maxwell before enrolling in his courses, but he found the great man's lecturing style obscure, "paradoxical and allusive". He learned from those shortcomings and strove to be clear and precise in his own presentations. His hand rules were a product of that ambition.

The rules express very simply the relationship between electrical current, magnetic field and motion in generators (with the right hand) and electric motors (with the left). In both cases the thumb, first and second finger should be pointed at right-angles to each other:

- The **TH**umb points in the direction of **TH**rust or motion
- The **FI**rstfinger points in the direction of the magnetic **FI**eld (north to south)
- The **C**entre or se**C**ond finger points in the direction of the **C**urrent (positive to negative)

Ambrose Fleming balanced his university work with a healthy practical involvement in the electrical industry. He acted as consultant to both the Edison Electric Light Company and the Marconi Wireless Telegraph Company, experiencing at first hand and solving problems in the practical application of electricity. When Marconi successfully made the first transatlantic radio transmission in 1901, it was with a transmitter designed by Fleming.

Fleming had signed an agreement with the company that, if the transmission were successful, all credit would go to Guglielmo Marconi himself. Marconi simply thanked Fleming among others for their work on the power plant for the transmitter; and as if that were not galling enough, Marconi, who made a career from capitalizing on other people's ideas, forgot that he had promised Fleming 500 shares in the company. It would have made the engineer a rich man. Fleming silently held his resentment until Marconi's death in 1937, when he finally felt able to speak out.

Nevertheless he continued to work with the company. In the course of attempting to improve long-distance radio reception he invented the diode, which became known as the Fleming Valve and was fitted in radar and radio equipment for the next fifty years until the development of solid-state electronics.

This small innovation made the spread of commercial radio broadcasting a reality. The diode led to the invention of the triode by Lee de Forest in America, and it is this form of amplifying vacuum tube of which rock guitarists speak when they extol the virtues of playing through a valve amplifier.

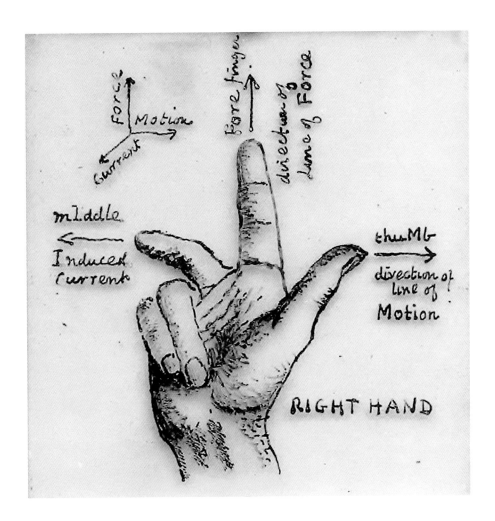

ABOVE: *Fleming's diagram is held at University College London (pictured left), where he taught electrical engineering. Fleming's rule helped his students determine the direction of a magnetic field, the motion of the conductor, and the resulting electromotive force.*

OPPOSITE: *Fleming worked for both Thomas Edison and Marconi, for whom he designed the radio transmitter with which the first transatlantic radio transmission was made.*

# Heinrich Hertz

## (1857–1894)

## Radio waves

The image of research scientists as theory-nerds detached from the practicalities of the outside world is usually an unfair stereotype. But the man who discovered radio waves could see no application for them in the real world and declared them to be "of no use whatsoever".

Today we are completely dependent on radio waves in everything from television to smartphones. Radio waves have a central role in scientific enquiry and the exploration of deep space. But when Heinrich Hertz was asked about what his discovery would be used for, he replied, "Nothing, I guess."

Heinrich Hertz was born in Hamburg and showed early aptitude for science. After studies in Leipzig and Munich he undertook his PhD research in Berlin. It was his tutor there, Hermann von Helmholtz, who suggested he focus on the work of the Scottish physicist James Clerk Maxwell.

Maxwell was a mathematical physicist who proposed the theoretical existence of electromagnetic waves. His theory, offered in 1864, was supported by rigorous mathematics, which became known as Maxwell's Equations; but no one had been able to prove him right.

Helmholtz thought Hertz was up to the task but Hertz couldn't imagine what equipment he might construct which would deliver such a proof. Instead he put in work on electromagnetic induction, although he did check Maxwell's Equations and found them to be mathematically plausible.

Seven years later Hertz was appointed a professor at Karlsruhe University and experimented with an induction coil composed of a pair of conductors. He noticed that a high voltage discharge into one of the conductors created an unexpected spark in the other. This transmission suggested a starting point for the task that Helmholtz had originally set him.

Over the next three years Hertz corresponded regularly with his old tutor and conducted experiments. He produced electromagnetic waves which were measurable and conformed to Maxwell's Equations. They moved through the air in the same way that light does, confirming Maxwell's theories that light and heat were both forms of electromagnetic radiation.

Hertz was pleased to have discovered these phenomena, which became known as Hertzian waves for several years until the term "radio waves" became more widely used; but he was unimpressed by them. "This is just an experiment that proves Maestro Maxwell was right," he said. "We just have these mysterious electromagnetic waves that we cannot see with the naked eye. But they are there."

Other scientists however pounced on the discovery. Marconi and others saw the potential for wireless communication, first in person-to-person radio, then radio broadcasting, and eventually television, telephone and satellite networks. Modern life depends on Heinrich Hertz's discovery; and his name is preserved in the international unit the Hertz, which refers to the frequency per second of any given event or wave.

# Friedrich Reinitzer

## (1857–1927)

## Liquid crystals

Everyone knows there are three possible states for matter – frozen, melted and boiled – for example ice, water and steam. Austrian botanist Friedrich Reinitzer was baffled, therefore, when he discovered that some compounds have *two* melting points, between which they exhibit properties of both solid and liquid.

Reinitzer was born in Prague, then a city of the Austrian Empire, and studied chemistry. He was particularly interested in the chemistry of plants and spent time studying the properties of cholesterol and compounds derived from it.

Cholesteryl benzoate behaved in a most unexpected way. As he heated it to 145.5°C (293.9°F) it melted from its solid state into a cloudy liquid; but as he further raised the temperature (to 178.5°C/353.3°F) it underwent a further change of state, a second melting, into a clear liquid.

Other chemical botanists had noticed the progressive change in appearance of cholesterol derivatives during heating; but none had thought it anything other than a visual effect of the change in temperature. To his credit Reinitzer wanted to know more. He studied the benzoate in its condition between the two melting points and noticed not only that it was iridescent but that it was birefractive – that is, it produced two images when light was shone through it.

Reinitzer knew enough to deduce that this in-between state must therefore have a different structure compared to the liquid of the higher melt. He corresponded with a German crystallographer, Otto Lehmann on the subject and they compared samples. Lehmann confirmed through a microscope that this intermediate state seemed to be crystalline, despite being liquid. Reinitzer presented his findings in a paper to the Vienna Chemical Society in 1888 including the unusual optical properties of the new state under polarized light.

And there he left it. The Austrian botanist did not pursue his discovery, which he saw as being of no botanical interest. Lehmann however saw it as something much closer to his own heart and pursued the phenomenon. The crystals that he had seen through a microscope suggested it was solid, but it flowed like a liquid. It was he who coined the term "liquid crystal" in 1904.

He and another German scientist, Daniel Vorländer, conducted thorough experiments into the nature of liquid crystals, synthesizing many in the laboratory. Although the discovery and their investigations aroused considerable interest at the time, liquid crystals were perceived as no more than a novel curiosity. At a time when atomic physics was at the cutting edge of scientific research, no one could see a practical use for them, and interest in them waned.

Only in 1958, when the world was well into the atomic age, was curiosity about liquid crystals rekindled. That year, Glen Brown of the University of Cincinnati published an article about them; and in 1965 he organized a conference which accelerated research into them with a view to practical applications. Electronics company RCA, the Radio Corporation of America, was one of the first to consider their use in flat displays after their research scientist Richard Williams noted the effect of passing an electrical charge through a thin layer of liquid crystals.

Today liquid crystal displays, LCDs, are central to our everyday lives in everything from wristwatches to computer screens. Liquid crystals can also be found in detergents, and in some proteins and cell membranes, a reminder that it was a botanist who first discovered their existence.

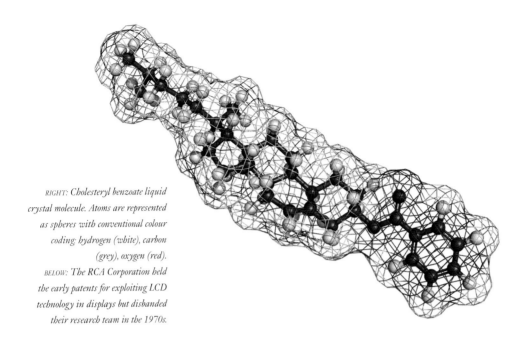

*RIGHT: Cholesteryl benzoate liquid crystal molecule. Atoms are represented as spheres with conventional colour coding: hydrogen (white), carbon (grey), oxygen (red).*
*BELOW: The RCA Corporation held the early patents for exploiting LCD technology in displays but disbanded their research team in the 1970s.*

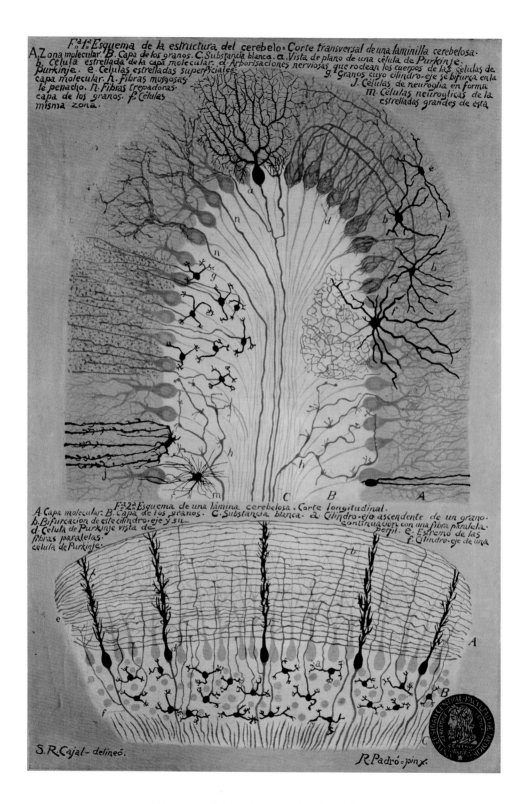

# Santiago Ramón y Cajal

## (1852–1934)

## Neuron theory

Passed from school to school for bad behaviour, an eleven-year-old Ramón y Cajal once blew his neighbour's gate off with a homemade cannon and received a prison sentence for his act. The wild child grew up to be a pioneer in the neuropathy of the brain and the first Spaniard to receive a Nobel Prize.

The young Ramón wanted to be an artist, to the great disappointment of his surgeon father. The latter finally persuaded his son to go to medical school by taking him to the cemetery to draw old bones. After his training Cajal held a succession of posts at the universities of Zaragoza and Valencia, developing interests in cellular microbiology.

Following an appointment as professor at Barcelona University he came across the work of Camillo Golgi, an Italian physician who had devised a method of staining nerve cells – neurons – in order to study them. Cajal was intrigued.

Neurology was still in an early stage of development. René Descartes (1596–1650) had speculated that human responses to stimulation were the result of a nervous system; and Luigi Galvani (1737–1798) had demonstrated that muscle contraction could be stimulated by electrical charge. Czech anatomist Johann Purkinje (1787–1869) conducted pioneering work on cells and was the first to describe neurons. It was assumed that the nervous system, like the circulatory system, was a network of continuous nerves – the reticular theory of neurobiology.

Golgi's stains made it possible to identify neurons in the brain for the first time, and were therefore a breakthrough in medical research. Unfortunately for Golgi, he subscribed to the reticular theory and saw all his observations as a reinforcement of that.

Cajal experimented with Golgi's stains and found a way to make them stick to more cells and therefore yield more research material. He also used deeper tissue samples under the microscope. The results were conclusive: Cajal could see that the neurons were separate from each other – contiguous, but not continuous. He could also see projections on the neurons, growth cones from which neurons send out synapses to communicate with other nerve cells, the first time these had ever been observed.

Cajal's discoveries about nerve cells, later named neuron theory, mark the start of modern neuroscience. With a proper understanding of how the nervous system works, it became possible to address the causes of failures in the system, and perhaps even to repair them. Cajal's improved staining techniques have proved valuable in diagnosing brain tumours, and the young artist Ramón would be delighted to know that the drawings of neurons with their tree-like growth, which he made as an adult, are still used in the teaching of neurobiology.

Golgi disagreed completely with Cajal's findings. It was therefore a surprising decision by the Nobel committee to award the 1906 prize for physiology jointly to Golgi and Cajal. Since the two men supported the opposing reticular and neuron theories, one of them was bound to be wrong. There is, however, no doubt that Cajal would never have been able to make his neural discoveries without Golgi's techniques, something which he acknowledged in his autobiography.

*OPPOSITE: The microscopic anatomy of the cerebellum drawn by Cajal and enhanced by R. Padro of the School of Medicine, Madrid.*

# Dmitri Ivanovsky

## (1864–1920)

## Viruses

As 2020's coronavirus pandemic demonstrated all too well, viruses can spread with deadly speed. Yet less than 150 years ago the very existence of viruses was unknown. We owe their discovery to a disease of tobacco plants.

Viruses are tiny particles of genetic material coated in protein. Bacteria, by contrast, are microscopic organisms, first discovered in the seventeenth century. Our understanding of their role in disease grew in the following centuries, particularly in the wake of the famous outbreak of cholera in London's Broad Street in 1854. Physician John Snow halted the spread of the disease there by preventing people from drawing water from the contaminated Broad Street water pump. Some water companies were already filtering their supplies by then, and others began to follow suit.

Dmitri Ivanovsky, a Russian botanist from St Petersburg, was dispatched to Bessarabia and the Ukraine in 1887 and the Crimea in 1890 to investigate a plant disease that was ravaging the economically important tobacco plantations of the region. The leaves, the most useful part of the plants, became mottled pale and dark green; and the growth of the plants was stunted.

Bacteria were suspected, and it was known that nearby plants infected each other. In an attempt to isolate the bacterium, he made a solution of infected leaves and passed it through a very fine porcelain filter, recently invented by the French microbiologist Charles Chamberland, which was capable of trapping bacteria.

To Ivanovsky's surprise the filtered solution was just as infectious to tobacco plants as the unfiltered preparation. He knew then that he had discovered a new mechanism for the transmission of disease. Today we know that Ivanovsky had found a virus; but he thought only that it must be a new form of micro-bacterium.

Ivanovsky wrote up his findings but did not pursue them. Eight years later a Dutch microbiologist, Martinus Beijerinck, repeated Ivanovsky's experiments and concluded that the agent of infection was not bacterial but something completely new: and it was Beijerinck who coined the term "virus".

Tobacco mosaic virus, the virus which Ivanovsky discovered, was also the first virus to be crystallized and its structure analysed by electron microscopy. DNA pioneer Rosalind Franklin published its structure in 1955.

More than any other frontier of science, virology is a battle of wits. The smallpox virus was eliminated in 1979 by an intensive programme of vaccination; but viruses evolve into more or less virulent strains. New viruses emerge, for example Ebola in 1976, Human Immunodeficiency Virus (HIV) in 1981 and Coronavirus 19 in 2020. The speed with which science has been able to develop a Covid-19 vaccine stems from Ivanovsky's and Beijerinck's work in the 1890s.

*ABOVE: The tell-tale effects of the tobacco mosaic virus
on broad tobacco leaves.*

*OPPOSITE: Ivanovsky was celebrated in his native Russia in 1964,
the centenary of his birth in Nizy, near St Petersburg.*

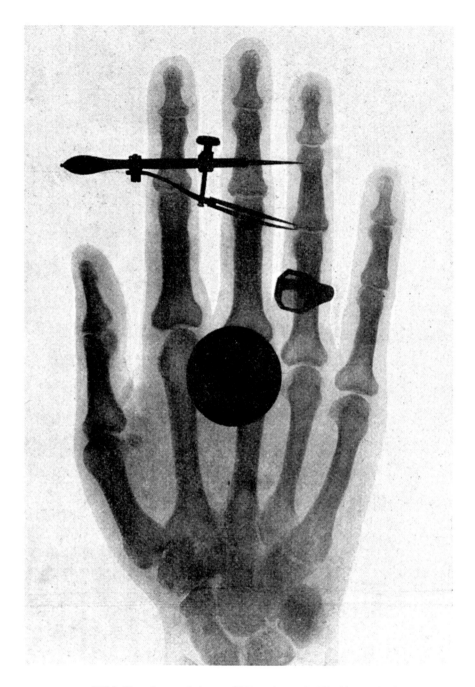

# Wilhelm Conrad Röntgen

## (1845–1923)

## X-rays

Some discoveries are pursued, while others are stumbled upon by chance. But as the Roman philosopher Seneca put it, "good fortune occurs at the meeting of opportunity with preparation". Wilhelm Röntgen's open-minded research of one phenomenon meant he was prepared for the discovery of another.

Wilhelm Röntgen's education was disrupted by relocation, expulsion and under-qualification. Yet he demonstrated perseverance and determination; he graduated in Zurich having studied under Rudolf Clausius, thermodynamics pioneer, and August Kundt, who experimented with light and sound in different gases. Kundt recognized Röntgen's commitment and hired him as an assistant at his new post in Würzburg University.

Following a string of lectureships and professorship in other universities, Röntgen returned to occupy the chair in physics at Würzburg in 1888. There he pursued an interest inherited from Kundt in the behaviour of gases. Röntgen was curious, like Kundt, about the refraction of light in different circumstances, including the effect on polarized light of electromagnetic fields. At a time when electricity was still little understood, the field was wide open for new research.

Röntgen was experimenting in late 1895 with a Hittorf-Crookes glass vacuum tube and a Ruhmkorff induction coil that could discharge high-tension electricity into the tube. The current passing through a filament in the tube (much like that of a light bulb) was already known as a cathode ray, and Röntgen was hoping to observe its fluorescent effect.

Unexpectedly, the fluorescence showed up several metres away from the tube on a piece of

equipment coated in barium platinocyanide. This suggested to Röntgen that previously unknown rays were at work. Because they were invisible he assumed that they were not light but some new kind of emission altogether; and being unknown he called them X-rays. We now know that X-rays are, like light, a form of electromagnetic radiation but of a higher frequency than visible light.

Röntgen experimented with different enclosures for the tube and found that some materials blocked more of the X-rays than others. To the horror of his wife, when he held her hand still between the tube and a photographic plate it recorded an image of her skeleton, the bones being less transparent to X-rays than the flesh surrounding it. He had not only taken the first X-ray photograph but also identified the great medical use to which such images could be put.

The rays came to be known formally as Röntgen rays, and the photographs as röntgenograms. As a diagnostic tool for medical science, they transformed the treatment of internal physical ailments, which could previously only have been seen by speculative surgery on the patient. Röntgen was awarded the very first Nobel Prize for physics in 1901 for his discovery. Although X-ray images are limited in their detail compared to the more sophisticated scanning devices of recent years, they remain a valuable aid to the work of saving lives all over the world.

# Svante Arrhenius
## (1859–1927)
## Link between $CO_2$ and global temperature

Svante Arrhenius was a polymath, trained in both chemistry and physics, who made significant scientific contributions in those and other fields. It is for his prescience in the matter of global warming that many remember him today.

His early work was on the behaviour of electrolytes – solutions that can conduct electricity. His dissertation at Uppsala University in 1884 contained the seeds of the discovery that would later earn him a Nobel Prize, that electrolytes separate into charged ions in solution even when no current is present.

His supervisors at the university were unimpressed with his thesis, partly because he had rejected them as dull and rigid in their thinking and opted to be taught instead at the Swedish Academy of Sciences in Stockholm. Arrhenius sent copies of his thesis to more open-minded scientists around Europe, who were much more encouraging.

He received his Nobel Prize in 1903, the first Swede to be rewarded by the Nobel Institute. Arrhenius had been instrumental in setting up the Institute in 1900 and sat on the Nobel physics and chemistry committees for the rest of his life. He insisted that nominations should be received from abroad, not only from Scandinavian scientists, thereby ensuring that the awards would be international and globally important. He was, it must be said, not above nominating his friends and blocking the nominations of his enemies on occasion.

In his later career he turned to biochemistry, studying immunology and the role of toxins and anti-toxins. He was also curious, in the cold climate of Sweden, about the causes of ice ages.

French physicist Joseph Fourier first introduced the idea, in the 1820s, of the Earth's atmosphere acting like the glass of a greenhouse to retain some of the heat received from the Sun. Although an oversimplification, the greenhouse analogy has stuck. In studies of infrared radiation Irish scientist John Tyndall was the first, in the 1860s, to consider precisely which gases in the atmosphere were most effective in trapping heat. Of the so-called greenhouse gases that he found, $CO_2$ was able to absorb a broad band of wavelengths of light, confirming earlier speculation by French physicist and mathematician Claude Pouillet that water vapour and $CO_2$ were the principle agents.

Arrhenius cited Pouillet's work in his own calculations. He estimated how much heat was retained by greenhouse gases through infrared observations of the Moon, and found that halving the $CO_2$ in the atmosphere would cause a drop in global temperatures of around 5°C (41°F). This would be enough to cause an ice age; and it encouraged Arrhenius to calculate the effect of a comparable rise in $CO_2$. It was the same: a rise in temperature of 5°C for a doubling of the gas.

Arrhenius went further and predicted in 1896 that $CO_2$ in the atmosphere would double in 500 years because of the industrial age's reliance on fossil fuels for power. He thought that global warming would be a good thing for the Swedish climate. We know differently today, but his theory and his calculations have been rigorously tested and found good. Today they lie at the heart of climate science.

*OPPOSITE AND ABOVE: Svante Arrhenius, a Swede who thought global warming would be a good thing, and a mural of Greta Thunberg in Bristol, a Swede who knows it is a catastrophic thing.*

*ABOVE: Marie Curie was the first person to win or share two Nobel Prizes. A delegation of Polish intellectuals encouraged her to return to Poland and continue her research in her native country. Instead she persuaded the French government to support her wish to create the Radium Institute.*

*LEFT: Pierre Curie and Maria Skłodowska. After completing her studies in Paris, Maria returned to her native Poland and Pierre wrote her a letter, begging her to return.*

# Marie Curie

## (1867–1934)

## Theory of radioactivity and discovery of polonium and radium

Without doubt the most famous woman in science, Marie Curie's whole life was dedicated to research into radioactivity, and she died because of it. The benefits of her work have saved the lives of millions.

Marie Curie was born Maria Skłodowska in Warsaw. Her mother ran a boarding school and her father taught physics. Both were passionate Polish nationalists at a time when the country was occupied by Russia. Throughout her life Marie retained her Polish identity: it is no accident that she named a newly discovered element after the land of her birth.

Her gender prevented her from studying science officially in Poland, although she learnt from her father and from the underground "Floating University" organized by Polish patriots to preserve Polish culture. She read every scientific book she could find, and acquired some practical experience working in the laboratories of the Museum of Industry and Agriculture in Warsaw. She worked as a governess to earn the money to study in Paris, and moved there in 1891. There she shared a laboratory with a young physics researcher, Pierre Curie, and the couple married in 1895.

Marie Curie's imagination was captured by new developments at the frontier of science. Within a few months of her marriage, German physicist Wilhelm Röntgen discovered X-rays; and the following year Frenchman Henri Becquerel detected radiation from uranium compounds, which he named "uranium rays". Here, she decided, was an interesting field of enquiry for her PhD.

Working with a uranium mineral called pitchblende, she found that the radiation was even stronger than from uranium alone. The mineral must contain another, more powerfully radiating element. She coined the term "radioactive" and began to examine other elements for radioactivity.

In her analysis of pitchblende she discovered two entirely new elements, which although present in minute quantities, were the stronger radioactive components of the mineral. She named the first polonium, from "Pologne", the French word for her native country. The second she called radium, from the Latin word for a ray; a metric tonne of pitchblende contains just 0.1g of radium chloride.

For their work on radioactivity Pierre and Marie Curie shared the 1903 Nobel Prize for physics with Henri Becquerel. It would take her another seven years to separate pure radium from the chloride; and in 1911 she received the Nobel Prize for chemistry for her discovery of radium and polonium. She was the first person to receive two Nobel prizes and remains one of only two to earn them in different scientific disciplines.

Despite her achievements, as a woman, she was not allowed to present her own findings at a lecture to the Royal Institution in London in 1903 – Pierre had to read them out – and the French Academy of Sciences felt unable to elect her or any woman to membership until 1962, when one of Curie's former students received the honour.

The Curies noticed the harmful effects of radium on skin cells and its ability to kill cancerous cells faster than healthy ones. But they did not fully appreciate the internal damage that it could do. Marie took no precautions to shield herself when, throughout World War I, she helped treat over a million injured soldiers on the front lines with mobile X-ray units, known as "petites Curies". She died from the cumulative effects of a lifetime's radiation in 1934. Some of her notebooks remain radioactive.

# Karl Landsteiner

## (1868–1943)

## Blood groups

Blood transfusions were risky interventions in the nineteenth century. Too often it resulted in hemagglutination – coagulation of the blood – which could be fatal. The danger was known but not understood, until Austrian immunologist Karl Landsteiner discovered the cause.

Karl Landsteiner trained as a doctor and became interested in immunology while still a student in Vienna. One of his student essays was about the effect of diet on the blood; and instead of practicing medicine after receiving his doctorate he studied chemistry in Germany and Switzerland.

He returned to Vienna in 1893 to work in the city's Pathological-Anatomical Institute. He is said to have conducted over 3,600 autopsies in the following ten years, and he wrote prolifically on many aspects of physiology including bacteria and viruses. He made a particular study of blood serum, the yellowish fluid that carries human blood around the circulatory system.

He found that when he combined the blood of two people, and agglutination occurred, it was because one person's blood came into contact with the other's serum. At the time, agglutination was believed to be caused by disease; but Landsteiner knew that both his samples were from healthy individuals.

He experimented with further samples from other people and found that some combined without agglutinating, and deduced that not all human blood was the same. In a series of tests he discovered the four most common blood groups, which we now call A, B, AB and O. When hemagglutination occurred, it was because one group was incompatible with another and carried group-specific antigens that the receiving serum saw as hostile, triggering hemagglutination as an immune response.

Landsteiner also worked out the pecking order of blood groups: type O humans can donate blood to

anyone but only receive type O; type A can receive from O but not B, and can donate to AB (and vice versa for type B); and type AB can receive any other type but cannot safely donate to anyone other than type AB.

The first blood transfusion made according to Landsteiner's findings was performed in New York's Mount Sinai Hospital in 1907. He received the 1930 Nobel Prize for physiology in recognition of his life-saving discovery of blood groups. Besides the obvious advantage of this knowledge for patients, Landsteiner also foresaw its application in the crime lab, where identifying the blood type on a piece of evidence could help narrow the field of suspects. With Alexander Wiener he went on to discover the Rhesus factor in blood in 1937.

If Karl Landsteiner had never conducted research into hemagglutination he would still deserve a place in the immunologists' hall of fame. He said himself that the discoveries of blood groups were just luck and that anyone could have made them. In conjunction with his countryman Erwin Popper he identified the polio virus, which led to Jonas Salk's development of a polio vaccine in the 1950s. Working with Viktor Mucha's new technique of dark-field microscopy he also identified the bacteria which cause syphilis.

The discovery of which he was most proud was that of haptens, small molecules that provoke no immune response unless combined with larger proteins. The use of synthetic haptens greatly expanded the possibilities of immunochemical research, with benefits for all mankind.

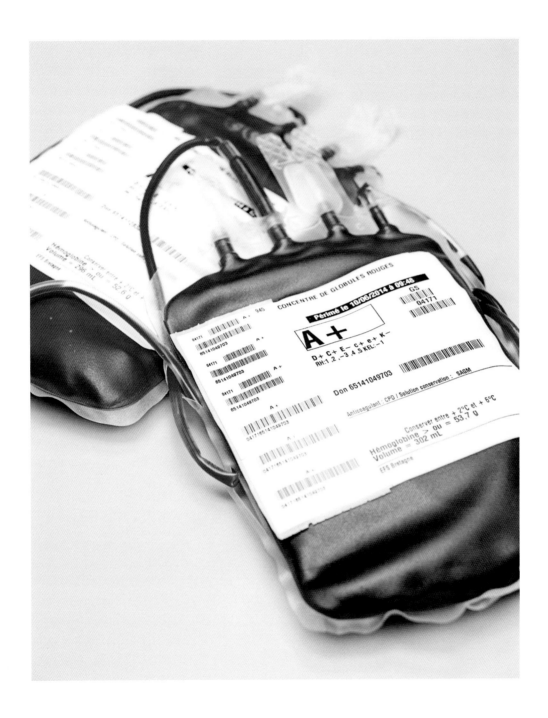

Label on blood pack:

CONCENTRE DE GLOBULES ROUGES
Périmé le 10/08/2014 à 09:48
GS
04171

**A +**

D+ C+ E− C+ e+ K−
RH:1 2 −3 4 5 KEL:−1

Don 65141049703

Anticoagulant : CPD / Solution conservation : SAGM

Conserver entre + 2°C et + 6°C
Hémoglobine > ou = 53,7 g
Volume = 302 mL
EFS Bretagne

OPPOSITE: *An Austrian stamp from 1968, the centenary of the birth of Karl Landsteiner.*
ABOVE: *Blood packs ready for transfusion with the all-important blood group*
*prominently displayed.*

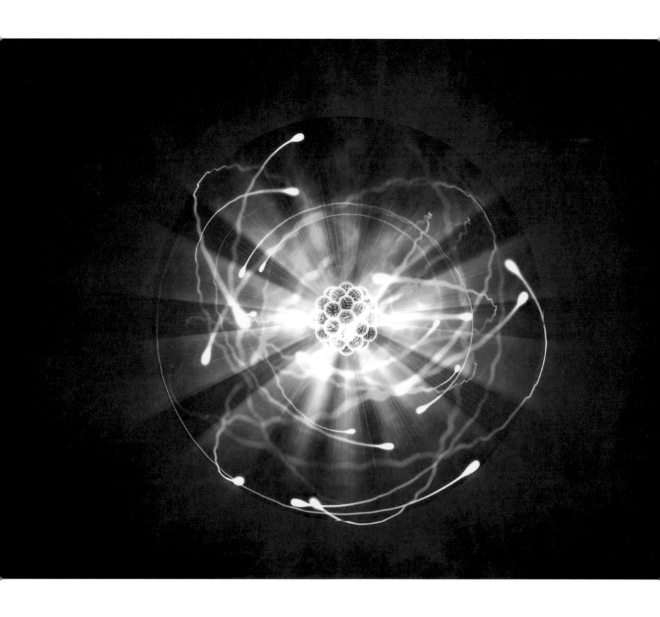

*ABOVE: Quantum theory helps explain the behaviour of subatomic
particles and how they interact.*

# Max Planck & Wolfgang Pauli

## (1858–1947 / 1900–1958)

## Quantum theory

Classical physics, the laws that had been stated and refined over thousands of years, can no longer explain everything that we observe. Modern physics is underpinned by two new theories – the general theory of relativity and the bafflingly difficult concept of quantum theory.

Richard Feynman, who won the 1965 Nobel Prize for physics, once admitted, "I think I can safely say that nobody understands quantum mechanics." Niels Bohr, Nobel physics laureate in 1922, went so far as to suggest that it didn't matter whether the elementary particles of quantum mechanics actually existed or not. It was enough that they provided a way of understanding the properties of matter – in other words that quarks, photons and so on were simply useful metaphors. Quantum theory is as much a question of philosophy, therefore, as of science.

Quantum theory has been simplistically described as the physics of very small things, while the general theory of relativity applies to the very large – the objects in the universe. Since its inception at the start of the twentieth century, quantum physics has been adapted, expanded and honed to explain more and more of the physical world, including its chemistry. Many great names of science have contributed to its development, but two stand out – Planck and Pauli.

In the last years of the nineteenth century, German theoretical physicist Max Planck began to consider a question first posed in 1859 by his fellow German, Gustav Kirchhoff. Why does electromagnetic radiation from a black body (an idealized physics concept which absorbs all frequencies of radiation) vary in intensity depending on the frequency of the radiated light and the temperature of the body? Another form of the question might be, why do hot black coals glow red?

Planck was an old-school believer in classical physics, and was forced almost against his will to the conclusion in 1900 that electromagnetic energy could only be radiated in whole number multiples of a basic amount – just as, on a staircase, you can't stand on anything between the steps. Planck called this minimum amount a quantum. The higher the staircase, the more quanta there are, but the step is the elementary element, the quantum. For this leap of faith, Planck is regarded as the founder of quantum theory, the theory of minimum amounts, smallest particles and least steps.

Planck spent many years trying to integrate quantum theory and classical physics, but the two were incompatible. Einstein confirmed Planck's theory in 1905 when he identified photons as the elementary particles, the quanta, of light. Niels Bohr proposed in 1913 that electrons, another kind of elementary particle, could only accumulate energy in quantum leaps, not in gradual increments. This implied that they circled the nucleus of an atom in one of several fixed orbits, between which they jump depending on their energy levels.

Austrian physicist Wolfgang Pauli made one of the most important discoveries of the quantum era when, in 1925, he found that identical fermions, a classification of particles that includes the electrons in atoms, could not co-exist in the same space. The Pauli Exclusion Principle explains through quantum mechanics why our everyday world is stable and solid and why the atoms of certain elements are more likely to combine with those of others to make compounds. It accounts for the mechanical, electrical, magnetic, optical and chemical properties of solid objects.

It is also the reason for the formation of white dwarfs and neutron stars and the formation of black holes; so quantum physics is not just for the little things in life.

# Ernest Starling & William Bayliss

## (1866–1927 / 1860–1924)

## The first hormone, secretin

William Bayliss and Ernest Starling were long-time friends and research colleagues. They became even more closely connected when Bayliss married Starling's sister Gertrude in 1893. By then the men had been collaborating on physiological studies for three years and would soon make a discovery of enormous importance.

Physiology is the study of how the body works, and medicine is the science of how to mend it when it doesn't. Starling had plans to become a London doctor until he spent a summer in the laboratories of German physiologist Wilhelm Kühne. Kühne studied the processes of vision, muscles and nerves, and latterly the digestive system – he discovered trypsin and invented the word "enzyme". Starling decided then and there to become a physiologist, and began to conduct research in the course of his work at Guy's Hospital in London.

Bayliss too decided to switch to physiology, having failed his anatomy exam at medical school. He taught at University College London. When their shared interest brought Bayliss and Starling into contact, they began to collaborate, first on the electricity of the heart and later on pressure in the veins and capillaries of the circulatory system.

They changed their focus to the digestive system in 1897, no doubt thanks to the enthusiasm shown by Starling for Kühne's work. Their collaboration was made easier when Starling was appointed professor at the university in 1899, and that year he demonstrated that the presence of food in the intestine triggers a nervous reaction which causes muscles to move the food through it, a process called peristalsis.

Next they turned their attention to the secretions of the pancreas that accompany food as it leaves the stomach for the intestines. The Russian physiologist Ivan Pavlov (he of Pavlov's Dog) believed that the secretions were triggered, liked peristalsis, by the nervous system when it sensed food. Starling and Bayliss conducted an experiment which disproved Pavlov's theory: even after they severed the nerves in the intestines of an animal, food passing from stomach to intestine triggered the same pancreatic secretions.

If the process was not sensory, it must be chemical. The pair discovered in 1902 that in the presence of food the wall of the intestine secreted something into the blood, which sent a signal to the pancreas. They gave this chemical messenger from the intestine a name – secretin.

It was the first such messenger discovered, and in 1905 Starling also gave the messengers a name: hormones. The existence of such messengers had first been suspected by Arnold Berthold, another German physiologist. His observations on the effect of castration on roosters hinted at the existence of a chemical later identified as the hormone testosterone. As we now know, hormones are a key production of the body, secreted in one part and circulated through the blood to regulate other parts. Our understanding of them is crucial in treating deficiencies either in the production of hormones or in their effectiveness.

Bayliss went on to study the mechanism of shock in wounded soldiers in World War I. Starling was instrumental in the establishment of teaching hospitals in Britain, and during the war he ensured that food rationing was designed according to the nutritional needs of the population.

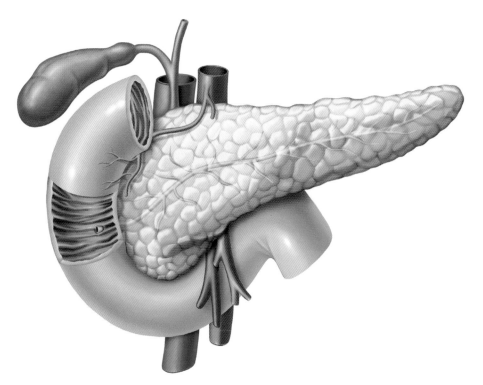

TOP: *Ernest Starling also created the Starling Equation, describing fluid shifts in the body, along with the mechanism of peristalsis and the discovery of secretin.*

ABOVE: *A diagram of the first part of the small intestine (duodenum) with the pancreas (in yellow) and the gall bladder (in green).*

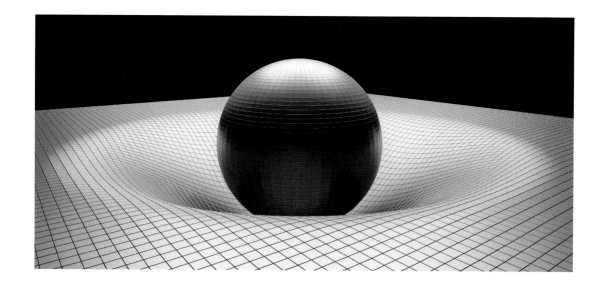

# Albert Einstein

## (1879–1955)

## Special and general relativity

Einstein's deceptively simple mathematical equation $E=mc^2$ defines the constant relationship between mass and energy anywhere in the universe. He published it in 1905 as part of his Theory of Special Relativity, an insight that has influenced almost every area of science since then.

Einstein arrived at the theory in his efforts to resolve conflicts between Sir Isaac Newton's laws of motion and James Clerk Maxwell's laws of electricity and magnetism. The theory attempted to encompass objects moving at extremely high speeds, near the speed of light, for which earlier laws did not work; and it proposed that at such speeds, distances appear shorter and time appears slower, warped by the masses of those objects.

The scientific concept of relativity is the principle that the laws of physics should apply in the same way in all admissible circumstances, called inertial frames. As an analogy, imagine standing in a speeding vehicle, and holding equal weights in each hand, one of which is held outside an open window. Now drop both weights. The weight dropped inside the vehicle will land at your feet; but the weight dropped outside the window will land several metres back down the road, with an apparently diagonal trajectory from hand to landing place. Inside the vehicle, hand, weight and landing place are in the same inertial frame.

Something similar was first proposed in Galileo's famous thought experiment set on board a ship, and efforts have been made ever since to find laws that would cover the situations of both weights. On an astronomical scale one sticking point was the concept of luminous ether, a theoretical medium in space through which light waves were supposed to travel, carrying light from its sources in the universe to us on Earth.

James Clerk Maxwell believed in the existence of luminous ether, but later in the nineteenth century scientists began to question the existence of a material that was infinite in its abundance but couldn't be seen and didn't interact with anything else. Experiments failed to find any trace of it in 1887, and in 1902 Dutch physicist Hendrik Lorenz showed that Maxwell's Equations of Electromagnetism could after all be applied universally without positing the existence of this ether, by some relatively simple mathematical transformations of the measurement of time and distance. Lorenz's work earned him the 1902 Nobel Prize for physics and paved the way for Albert Einstein's definitive theories of relativity.

The Theory of Special Relativity applies common rules to all those objects in the same inertial frame. The General Theory of Relativity, which Einstein proposed in 1915, works across all possible frames. Einstein reasoned that everything that happens, happens in a specific place and at a particular moment of time. He introduced the concept of spacetime, so that frames could be defined in four dimensions – three of linear measurement (space) and one of time.

Massive objects in space, the General Theory states, distort both space and time. It therefore has implications for our perception of absolutely everything, including gravity and light. And it was the launch pad for a brand new field of study, cosmology, which concerns not only the nature of the universe but its origins and the changes through which it has gone over time.

OPPOSITE TOP: *A 3D visualization of how gravity distorts electromagnetic waves close to planets.*
OPPOSITE BOTTOM: *Albert Einstein attributed his brilliant mind to having a childlike sense of humour.*

# Ernest Rutherford

## (1871–1937)

## Atomic nucleus

Ernest Rutherford is one of the true giants of science history. He is regarded as the Father of Nuclear Physics, a field of study which drove early twentieth-century research and led directly to the development of quantum theory.

He was blessed with a natural enthusiasm and an outgoing personality which inspired his co-workers. His instinct for the right way to approach a question and his willingness to experiment with answers earned him achievements in many areas of scientific activity. The breadth and depth of his contributions to science have led to comparisons with Isaac Newton and Michael Faraday.

Rutherford was awarded the Nobel Prize for chemistry in 1908 for his early research into radioactivity. While experimenting with thorium he discovered the concept of radioactive half-life – the length of time it takes for a radioactive element to decay. In the same research he identified and named three different kinds of radiation – alpha, beta and gamma rays.

He pursued these findings in the years following his Nobel award, exploring the nature and effect of these rays. His observations led him to the theory for which he is most famous – the Rutherford Model of the atom. In bombarding gold foil with alpha rays he observed that some alpha particles were deflected while others passed through. He concluded that the atoms of gold consisted of a small relatively heavy nucleus, which caused the deflection, surrounded by low-mass orbiting electrons through which the rays could pass.

This overturned John Dalton's idea that the atom was the smallest possible particle. If atoms consist of such sub-atomic particles as nuclei and electrons, then it must be, at least theoretically, divisible. Rutherford went even further and proposed the existence of protons, which along with neutrons we now know are the even smaller particles or nucleons of which a nucleus consists. (And on an even smaller scale, a proton consists of three quarks.)

For completely changing our understanding of the building blocks of the universe, Rutherford is often hailed as the greatest scientist in history. His ideas have led directly to the development of atomic energy and to an explanation of how elements are made. Among many tributes to him is the element rutherfordium, an element whose existence is only possible by laboratory synthesis.

Not only did Rutherford win a Nobel Prize; so did several of those with whom he worked, including J. J. Thomson (Physics, 1906) who taught him in the 1890s, Frederick Soddy (Chemistry, 1921), James Chadwick (Physics, 1932), Edward Appleton (Physics, 1947), Patrick Blackett (Physics, 1948) and John Cockcroft and Ernest Walton (Physics, 1951). In addition to his own achievements Rutherford was like a comet leaving a brilliant trail in his wake.

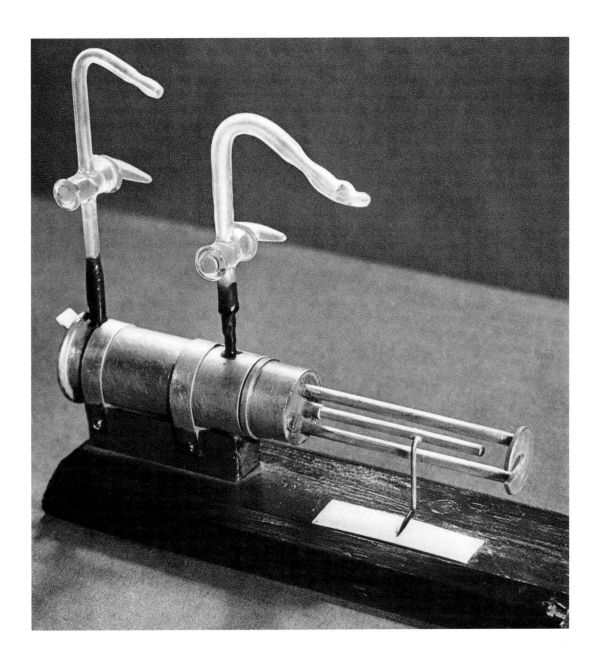

*ABOVE: Rutherford's apparatus with which he first observed artificial transmutation in 1919. In this deceptively simple set-up nitrogen atoms were converted into oxygen atoms when in collision with alpha particles from a source in the enclosed horizontal tube. Protons ejected by nitrogen when forming oxygen were detected at a rectangular window at the end of the tube.*
*OPPOSITE: New Zealander Rutherford influenced and inspired a generation of nuclear physicists.*

*ABOVE: A demonstration of Quantum Magnetic Levitation and the Suspension Effect. A splash of liquid nitrogen cools a ceramic superconductor, forcing it to float in the air below a magnet.*

# Heike Kamerlingh Onnes

## (1853–1926)

## Superconductivity

Electrical resistance, a measure of the resistance of a material or component in an electrical circuit to the flow of electricity, usually falls gradually with the lowering of temperature. Dutch physicist Heike Kamerlingh Onnes discovered some materials which behaved very differently.

Heike Kamerlingh Onnes received his doctorate from the University of Groningen in 1879, a year before another Dutch physicist, Johannes van der Waals (1837–1923), published important equations on the behaviour of gases. Van der Waals' equations were to be applied to *real* gases, and his work took into account the fact that molecules of different gases have different shapes and behave differently with each other. Hitherto, scientific laws had been modelled on so-called ideal gases – gases in which the molecules were regarded as simple points, moving randomly and not interacting.

Kamerlingh Onnes was fascinated and began to study the properties of gases in extreme conditions, especially at very low temperatures and in their liquid states. To this end, after he was appointed professor of experimental physics at the University of Leiden, he developed a cryogenic laboratory which became a world centre for this kind of research and which still exists today, bearing his name.

After more than a decade of continuous improvement in technique, Kamerlingh Onnes became the first person to liquefy helium, in 1908. Helium's boiling point is -269°C (4.2°K); and he managed to lower the liquid gas's temperature to -271.3°C (1.7°K), the lowest temperature ever achieved on Earth at the time. He was frustrated, however, in his attempts to solidify helium; that feat was finally achieved only five months after his death by his former student and his successor as director of the cryogenic facility, Willem Hendrik Keesom (1876–1956).

Kamerlingh Onnes continued his investigations with mercury, whose melting point of -38°C (234°K) was easier to achieve. When in 1911 he immersed a wire of solid mercury in liquid helium at -269°C and passed an electric current through it, he observed a remarkable event. Instead of declining *gradually* as it cooled, the resistance of the mercury wire fell *suddenly* to zero.

He repeated the experiment with other metals – tin and lead – and saw the same phenomenon. Kamerlingh Onnes understood the significance of this most unexpected result for the theory of conductivity in solids. A material with no resistance offers no inhibition to electricity; a loop of such material can carry an electric current forever, with no need for a power supply. Kamerlingh Onnes called such materials superconductors.

He was awarded the Nobel Prize for physics in 1913, the first of five such awards to scientists in the field of superconductivity. Subsequently, superconductors have been found which don't require such extreme low temperatures including, in 2020, one which can operate at room temperature. The potential applications for superconductors are still being explored. They make the most powerful electromagnets possible, and have found a use in MRI scanners, particle accelerators and magnetometers. They are useful in electricity generation, transformers and electrical motors.

One of the remarkable attributes of superconductivity is the complete ejection of magnetic field from within the superconductor, known as the Meissner Effect. The exterior magnetic field increases instead, and this has implications for the technology of maglev trains and other magnetic levitation applications. More than a century after Heike Kamerlingh Onnes discovered it, superconductivity is coming into its own.

# Alfred Wegener

## (1880–1930)

## Continental drift

In the Age of Empire, when all the continents were being mapped and colonized, geographers and others began to notice how well matched the east and west coasts of the Atlantic Ocean are, like pieces of a jigsaw. Was it mere coincidence?

Alfred Wegener, a German meteorologist, was neither the first to wonder why, nor even the first to suggest that the continents had once all been joined together. The prevailing view at the end of the nineteenth century was that the land between the present continental landmasses had simply subsided and been flooded by the oceans.

The geological concept of isostasy had emerged in the 1880s – the notion that the Earth's crust floats on a mantle

of molten lava, and that the height of the world's high places depends on the buoyancy and thickness of the crust on which they stand. The idea that land could suddenly sink was incompatible with isostasy, which implied equilibrium between the crust's buoyancy and the force of gravity.

Wegener began to look for an alternative explanation. If the crust is floating, then why should the original single landmass not have broken up like an ice floe and its fragments not have drifted apart? Wegener became convinced of the idea and studied the opposing coasts of the Atlantic Ocean to look for evidence that they were once joined.

He found many examples of the same sequence of geological strata occurring on both sides. He noted closely related fossil species recovered across several modern-day continents, suggesting that they had evolved while the landmasses were still connected. For example, the primitive fern *Glossopteris* is fossilized in the southern

regions of South America, Africa and India, as well as in Australia and Antarctica.

Wegener speculated on the cause of the drift, without reaching a conclusion. He thought it might be changes in the axis of the Earth's rotation; and he touched on the accurate notion of spread from the mid-ocean ridges, where the crust was torn open by lava welling up from below.

He first presented his theory of continental drift in 1912 and promoted it in publications and lectures for the rest of his life. But he was a meteorologist, not a geologist, and the scientific community that he had to convince were sceptical of his insights. He was an outsider. By the time of his death in 1930 few scientists were paying any attention to his theory.

Then, in the 1950s, new paleomagnetic studies could demonstrate changes on the magnetic polarity of ancient rocks. They proved, for example, that India had, as Wegener claimed, once been in the southern hemisphere. Geologists discovered in the 1960s that sea floors were spreading, just as Wegener had suggested. And with the invention of the Global Positioning System (GPS) in the 1970s it became possible at last to see and measure the extent of continental drift. Wegener was hailed as the prophet of a revolutionary understanding of the nature of our planet, the founding father of the science of plate tectonics.

*Polar researcher Alfred Wegener at the base camp for Johan Koch's 1912–13 Greenland expedition.*

*The Western hemisphere of the Earth during the Early Jurassic period. In this image the nascent North American continent has just broken away from North Africa while South America and the rest of Africa remain joined as Gondwana. To the west, in the global Panthalassa ocean, are strips of land corresponding to the Wrangellia Terrane which later merged with western North America.*

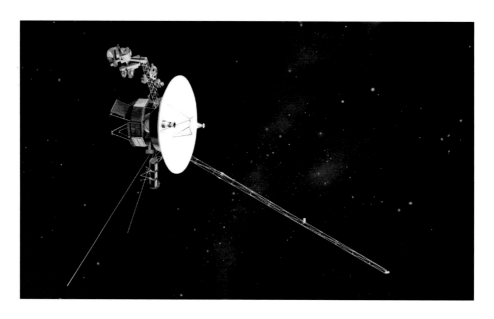

# Henry Moseley
## (1887–1915)

# Atomic numbers

Lay scientists know about the Periodic Table, the unsymmetrical chart of small squares with the symbols of all the elements. They may also have heard of atomic numbers, which match the order in which the elements are listed. But why are they in that (non-alphabetical) order?

Scientists love to make lists, and Russian chemist Dmitri Mendeleev made one in the form of the Periodic Table, based on each element's atomic weight – a relative figure which compares the mass of atoms in a given sample to those of carbon in a sample of the same size. He found that elements with similar chemical properties tended to be grouped together in the list, an interesting coincidence.

Ernest Rutherford, in his research on the nature of the atom, noticed another coincidence: a rough correlation between the electric charge in the nucleus of an element (in his case gold) and its position in the Periodic Table. A Dutch lawyer and amateur physicist Antonius van den Broek suggested that it might be no coincidence but an exact correspondence. However he lacked the resources to pursue the idea.

Henry Moseley decided to put the theory to the test. He was a physicist whose first job after graduation was as a demonstrator of physics experiments at Manchester University, where Rutherford was his supervisor. Rutherford offered him a fellowship at Manchester but Moseley preferred to return to Oxford University. There he conducted innovative experiments with many elements to observe their X-ray spectra and found a direct mathematical relationship between the wavelengths of the X-rays and the atomic weights of their elemental targets.

Although Mendeleev had ordered his Table according to weight and chemical properties, Moseley proved that these attributes were dictated by the nuclei – to be precise, that the number of protons in the nucleus of an element was the same as its position in Mendeleev's table.

This was not only a fascinating confirmation of the two earlier approaches to ordering the elements, Moseley's results also confirmed that there were gaps in the Periodic Table, which meant that there remained some elements yet to be discovered. In the half-century following Moseley's work, all of those elements were found, either in nature or by synthetic production in the laboratory. Their existence had not even been suspected in Moseley's lifetime.

Henry Moseley conducted his experiments in 1913. When World War I broke out in August 1914, he gave up his work and enlisted with the British Army's Royal Engineers. The following year he was posted to Gallipoli where he was shot and killed during a long and disastrous battle between forces of the British and Ottoman Empires. It has been suggested that, had he lived, he would have won the Nobel Prize for physics in 1916.

*ABOVE LEFT: In World War I Moseley was shot by a Turkish sniper while working as a technical officer on telephone communications . Isaac Asimov would later say "… his death might well have been the most costly single death of the War to mankind generally."*
*LEFT: Long-distance spacecraft, such as Voyager 1, rely on atomic batteries. Henry Moseley created the world's first atomic battery – a beta cell. He called it a radium battery.*

# Niels Bohr

## (1885–1962)

## Model of the atom

When Ernest Rutherford published his inspired model of the atom in 1911, it galvanized the atomic physics community. As other scientists looked at it through the prism of their own research, it proved accurate but incomplete. One, Danish scientist Niels Bohr, saw its implications for quantum theory.

Max Planck first proposed the quantum, the minimum amount of any agent in a reaction which was required to make the reaction work, in 1900. Planck applied the idea particularly to electromagnetic radiation, a field in which Niels Bohr completed his PhD in the summer of 1911. Bohr's research was in the electrons and magnetism of metal elements, and he had concluded that the latter could not be explained by the former alone.

In September of that year Bohr travelled to England to meet the great minds of that country's universities. He impressed Rutherford sufficiently to be invited to join him for a year in post-doctoral research at Manchester University. Bohr combined Rutherford's ideas with Max Planck's and proposed an atomic model in which electrons not only orbited around a neutron as Rutherford had suggested; they did so at different distances, like the planets around the Sun, and were capable of dropping from an outer orbit to an inner one by shedding a quantum of electromagnetic energy.

Rutherford was one of the first to applaud the Bohr Model. It was also enthusiastically received by the younger generation of nuclear physicists, including Albert Einstein, Max Born and Enrico Fermi. The model won admirers because it explained reactions and properties which other models, even Rutherford's, could

not. Better still, it could be used to predict the outcomes of experiments yet to be conducted; and the Bohr Model formed the basis of quantum theory for the next twelve years, until the development of quantum mechanics. He was awarded the Nobel Prize for physics in 1922.

The evolution of the atomic model in the twentieth century is a chain of cooperative endeavour from scientist to scientist. J. J. Thomson's model was superseded by his pupil Ernest Rutherford, whose model was improved on by his research assistant Niels Bohr. Bohr's model, too, was in time improved, first by German quantum physicist Arnold Sommerfeld and then by the Austro-German team of Erwin Schrödinger and Werner Heisenberg, who worked under Niels Bohr at the University of Copenhagen.

Niels Bohr taught at Copenhagen during World War I and after its conclusion he raised funds for an Institute of Theoretical Physics there. It opened in 1921 and still functions today, renamed the Niels Bohr Institute, as a forum for the exchange of ideas. During World War II he escaped to England and represented Great Britain in a mission to the US Manhattan Project. After that war he helped to set up CERN, the organization which supports European nuclear research. The highly radioactive synthetic element bohrium, first created by Soviet scientists in 1976, is named after him.

ABOVE: *The control room of Niels Bohr's Cyclotron.*
OPPOSITE: *Neils Bohr with Albert Einstein in the 1920s. Bohr, together with*
*Heisenberg, had views on quantum theory that conflicted with Einstein's own beliefs*
*and the pair were often involved in heated debates on the subject.*

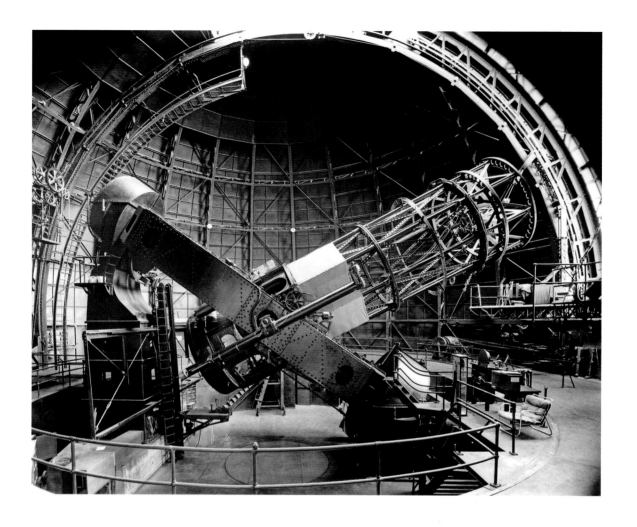

ABOVE: *A 1960s photo of the 100"
Hooker Telescope at the Mount
Wilson Observatory.*
LEFT: *The Andromeda Galaxy, also
known as Messier 31, and originally
the Andromeda Nebula, the nearest
major galaxy to the Milky Way.*

# Edwin Hubble

## (1889–1953)

## Galaxies beyond the Milky Way

It was once thought that our galaxy, the Milky Way, was the universe. Edwin Hubble discovered that far from being alone in the blackness of space, the Milky Way was just one of billions of galaxies, billions of light years apart and still moving further apart.

Like all great scientists Hubble's genius was in part his ability to see the implications of the discoveries of others, and to provide a new platform from which future scientists could make further discoveries. Hubble's greatest influence was perhaps his grandfather, a keen amateur astronomer who allowed young Edwin to look through one of his telescopes on the eve of the boy's eighth birthday. The child was so excited by what he saw that he pleaded to be allowed to stay up all night stargazing, even if it meant foregoing his birthday party the next day.

There was never any question about which direction Edwin Hubble's life would take, although for a while he also studied law and languages, which offered a more stable career choice. He earned his PhD in 1917 with a thesis entitled *Photographic Investigations of Faint Nebulae*, a study that laid the groundwork for his greater contributions to astronomy. He joined the staff of Mount Wilson Observatory two years later and remained there for the rest of his working life.

Mount Wilson benefits from the cleanest air in North America and is home to the massive Hooker 100" telescope, the largest in the world until 1949. Hubble continued his studies of nebulae, which were thought at the time to be clouds of dust and gas within the Milky Way. He became particularly interested in stars of a certain type, the cepheids, which were present in many nebulae.

Viewed from the Earth, cepheids appear to pulse, varying in luminosity and diameter, and American astronomer Henrietta Leavitt discovered a correlation between the brightness of cepheids and the frequency of the pulses. That correlation could be used to calculate the distance of a star from our planet at ranges of up to 20 million light years, an improvement for astronomers who were still using parallax and triangulation for measurements which were only possible for distances of up to a thousand light years. The Milky Way alone is 100,000 light years wide from side to side.

Hubble focused on the Andromeda Nebula and used Leavitt's discovery to calculate that it was some 900,000 light years away. It was therefore definitely not part of the Milky Way and, indeed, on closer inspection not a nebula either, but another galaxy. (With refinements and further observations, we now know that it is even further away at 2.48 million light years.) It emerged that many objects in space that were previously defined as nebulae were entire galaxies, millions of light years apart in a universe which was suddenly much bigger than anyone had suspected.

Hubble also built on the work of another American, Vesto Slipher, concerning red shifts in the visible spectrum of space objects. Slipher considered red shift as a version of the Doppler Effect. Just as a police car's siren falls in frequency as it gets further away, red shift (a change in the frequency of its light) denotes a star or galaxy that is getting further away from us. And Hubble found that *all* galaxies are moving away from us. Although he was reluctant to commit to the idea without further evidence, his observations are now seen as evidence of an expanding universe.

# Cecilia Payne-Gaposchkin

## (1900–1979)

## The composition of the Sun

Cecilia Payne was born in Buckinghamshire, England. Her father and aunt were musicians, and English composer Gustav Holst, who taught Cecilia at St Paul's School (and who, by coincidence, had just completed his orchestral suite *The Planets*), thought she showed musical promise.

She showed early ability in mathematics and was determined to follow a scientific path. She demanded a good schooling in mathematics and German (the lingua franca of the scientific community), and won a scholarship to the all-female Newnham College at Cambridge University, where she studied Physics and Chemistry. This was at a time when female students were not permitted to receive degrees, or even withdraw books from the library.

Payne's life was transformed by attending a lecture about photographing stars during a solar eclipse. The lecturer was Arthur Eddington, an early translator and supporter of Albert Einstein's General Theory of Relativity; his expedition to take photographs of the 1919 eclipse was one of the first proofs of the Theory.

Her awakening to the enormity, complexity and sheer beauty of the study of astronomy upturned her entire view of science. All too aware of the limited options available to her in Britain she became only the second person to take advantage of a new fellowship for women at Harvard University Observatory. She moved to Massachusetts in 1923 and remained there the rest of her life.

She was a brilliant student and she became the first woman to be awarded a PhD in astronomy by Radcliffe College, the female counterpart to Harvard's men-only student body. Through spectral studies of the elements present in the stars she discovered that, unlike the Earth, they were composed largely of helium and hydrogen. In her dissertation she claimed that hydrogen must be the most common element in the universe, and in our own star, the Sun – up to a million times more plentiful than the metallic elements.

This contradicted the received wisdom of the day, which held that heavenly bodies all had roughly similar compositions of elements, whether they were stars or planets. This view was shared by an old-school astronomer, Henry Norris Russell, who reviewed Payne's PhD; and he discouraged her from reaching such a definite and challenging conclusion, which he told her was "clearly impossible". Payne was correct in her deduction but at Russell's insistence she added that "the enormous abundances for these elements … are almost certainly not real."

Russian-American astronomer Otto Struve, however, praised her work as "the most brilliant PhD thesis ever written in astronomy" and, to his credit, Russell also acknowledged her genius when he reached the same conclusion as she had by different methods in 1929.

In other times Payne might have been awarded a Nobel Prize for her observations. Her studies of thousands of stars with her husband Sergei Gaposchkin are the foundation for all stellar research since. As it was, she was at first allocated only minor research projects at Harvard. In time, however, she became the first female professor in Harvard's Faculty of Arts and Sciences. Cecilia Payne-Gaposchkin not only revolutionized the discipline to which she dedicated her life. She remains an inspiration to women in *any* discipline.

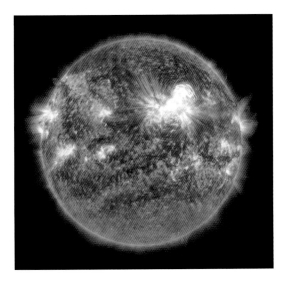

ABOVE: *Payne-Gaposchkin spent her entire academic life at Harvard University, but despite gaining the first PhD in Astronomy at Radcliffe College in 1925, she had to wait until 1956 for a professorial appointment.*
LEFT: *A NASA photo of the Sun showing solar flares.*
OPPOSITE: *The Harvard Observatory in the 1920s.*

# Erwin Schrödinger

## (1887–1961)

## The Schrödinger Equation

Is it a particle? Is it a wave? An early clash between classical and quantum physics in the understanding of how light travels was resolved by the discovery of two pillars of quantum mechanics – wave-particle duality and probability waves.

The elementary particles which form the basis of quantum theory are supposed to act individually. Photons are the elementary particles of light; yet we've known since the middle of the nineteenth century that electromagnetic radiation, of which light, heat, and other forms of energy are composed, behaves like a wave.

Imagine light hitting a pane of glass at an angle, causing (say) 70% of the light to pass through and 30% to be reflected from it. If light is a wave, the reflected light is simply a smaller wave. But if light is photon particles, each individual photon must have a 30% chance of not passing through the glass. This means an enormous number of possible outcomes for the light as a whole, including the very small possibility that none of the light is reflected, or that all of it is. Quantum theory thus implies that nothing is certain and everything is simply a matter of probability. It quite literally undermines the fabric of the classically understood universe.

French physicist Louis de Broglie demonstrated in 1924 that every particle, and therefore everything, has a wavelength, a condition known as wave-particle duality. Just as Max Planck had shown that waves of light could behave as particles, so de Broglie proved that particles could act as a wave. What drove this wave-like behaviour?

Erwin Schrödinger was an Austrian physicist who built on de Broglie's finding. In a leap of imagination, which so often in science is a prelude to discovery, Schrödinger proposed an imagined, insubstantial wave, which somehow informed the wave-like behaviour of particles – he called it a probability wave. Better still, he derived an equation, published in 1926, to predict the motion of this wave and therefore of the particles which it guides.

Schrödinger's Equation was a new and complete way to describe reality through the prism of quantum theory. As Max Born proved, it can be used to calculate the probability of a particle being at a particular place at a particular time, and it has become central to quantum mechanics. Because it has been shown to apply to all elementary particles and therefore all things made of particles (which is everything), it can be applied to anything from atoms to galaxies.

Schrödinger is the same man who is remembered by non-scientists for Schrödinger's Cat, a thought experiment designed to show the absurdity of a particular approach to quantum physics called the Copenhagen Interpretation. That interpretation accepts the concept from classical physics of wave superposition, in which waves can be combined, and applies it to quantum physics as quantum superposition. In the closed system of a steel box in which a cat has an equal probability of dying and living, Copenhagen says that both states exist, superimposed, until the box is open and one of the quantum states collapses. Until then, the cat is, according to the Copenhagen Interpretation, both alive and dead.

Schrödinger found the idea ridiculous. However, there are many other interpretations of quantum mechanics and of Schrödinger's experiment. His poor cat has become a test of comparative philosophy as much as one of quantum interpretation.

*OPPOSITE TOP: Erwin Schrödinger photographed in 1933.*
*OPPOSITE BOTTOM: Schrödinger's Cat – alive and dead at the same time – was the physicist's memorable illustration of why the Copenhagen Interpretation didn't make sense.*

# Werner Heisenberg

## (1901–1976)

## The Uncertainty Principle

The Uncertainty Principle (not to be confused with the Probability Wave) is further proof that, under quantum theory, nothing can ever be completely known. Of that, Werner Heisenberg, who discovered the principle, was absolutely certain.

A known issue in physics research is the Observer Effect, which occurs when the very act of conducting an experiment affects the outcome. An often quoted example is the checking of air pressure in a tyre: the attachment and removal of the pressure gauge always lets out a little air, thus changing the result.

It's often assumed that the Observer in this Effect is a person, but this is not true. A machine designed to monitor the temperature in a room, for example, might itself generate heat which distorts its own reading. Generally the Observer Effect can be mitigated or avoided altogether by small changes in experimental conditions – in this case by shielding the warm parts of the machine. Heisenberg's Uncertainty Principle, however, suggests that by its very nature the quantum universe can never be accurately observed and measured.

Heisenberg's work in the early days of quantum theory earned him the 1932 Nobel Prize for physics for his description of quantum mechanics, first outlined in his 1925 paper "Quantum theoretical re-interpretation of kinematic and mechanical relations". It was while pondering the observability of quantum mechanics that he came up against an insurmountable version of the Observer Effect.

You can't see electrons through an ordinary optical telescope because their wavelength is shorter than that of visible light. To mitigate this Observer Effect,

Heisenberg therefore imagined a microscope that used gamma rays, whose wavelengths are shorter than electrons, to "see" them. But at this subatomic level, gamma rays would alter the momentum and direction of electrons when they struck them, altering their observed behaviour. And, in fact, Heisenberg realized, photons of light would have the same effect on electrons even if you couldn't see it.

From these thought experiments he reasoned that it was impossible to know both elements of some pairs of variables. For example if the energy state of a given particle is known, its stability can't be. If its position is certain, its speed and direction cannot be. And in quantum systems there can be no mitigation of this observer effect. The more certain the value of one variable, the less certain that of the other. Heisenberg even devised an equation to show the relationship between the two variables, defined by a constant, the Planck constant.

The implication of the Uncertainty Principle is that quantum particles can never be fully described, but only exist in a quantum state. With it Heisenberg sounded the death knell for classical physics, whose whole approach was to discover the certainties of the natural universe. Now, it was clear, it could only deal in probabilities, and even then only up to the inevitable limits of the Uncertainty Principle.

*ABOVE: A Nobel Prize winner in 1932, Heisenberg was put in charge of the research programme to build a German nuclear weapon in World War II. However, thanks to Hitler's persecution of Jewish scientists and his politicization of academic institutions, many of the top physicists had already left the country.*

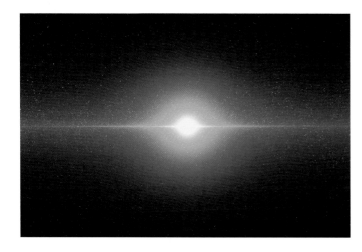

LEFT: *A computer representation of the Big Bang. The phrase "Big Bang" was used to deride Lemaître's theory, but he was proven correct.*

BELOW: *To celebrate the great Belgian scientist, the European Space Agency named an Automated Transfer Vehicle (ATV) for him. It was used to transport 6.6 tonnes of cargo to the International Space Station in 2014, and in this photo, is about to dock.*

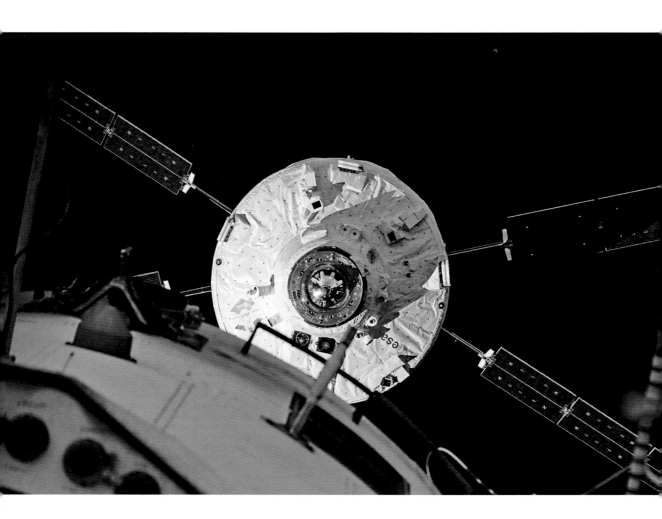

# Georges Lemaître

## (1894–1966)

## Big Bang Theory

In the pauses between the horrors of trench warfare in World War I, eighteen-year-old Private Georges Lemaître was able to block it all out and read books on physics. This was the mind that, years after the explosions of battle, would propose the greatest explosion of them all – the Big Bang Theory.

What made the universe? It's a question that only religion had dared to answer before the twentieth century. Creationists believe that a god made everything. It was, by a quirk of fate, a Catholic priest who first proposed that the cosmos began as a single point that exploded; and that the universe – the debris of that explosion – is still being flung outwards from that point.

Georges Lemaître, who was educated by Jesuits but trained for the church as a Roman Catholic, kept his religion and his physics completely separate. He studied both and in 1923 was both ordained as a priest in his native Belgium and employed as an astronomy research associate at Cambridge University in Britain. He continued his astronomy research at Harvard and MIT in the US.

Lemaître was not the first to consider an expanding universe. In response to Albert Einstein's general theory of relativity, Dutch astronomer Willem de Sitter, who later co-authored papers with Einstein, proposed a simplified model called the de Sitter Universe. Russian physicist Alexander Friedmann did work on Einstein's field equations which supported two models of the expanding universe – a "big bang" origin, and Steady State theory in which, as fast as the universe is expanding, new material is being produced.

Friedmann's ideas did not gain much traction because he was working in the earliest years of Communist Russia. Lemaître, too, was hardly noticed at first. He took up a professorship in astrophysics at the University of Leuven in Belgium in 1927 and in the same year he published his paper *Un Univers homogène de masse constante et de rayon croissant rendant compte de la vitesse radiale des nébuleuses extragalactiques* ("A homogeneous Universe of constant mass and growing radius accounting for the radial velocity of extragalactic nebulae"). It was printed in the *Annales de la Société Scientifique de Bruxelles*, a scientific journal with a small circulation.

That mouthful of a title contained Lemaître's first exposition of the Big Bang Theory, something many other scientists found hard to swallow. Einstein, who believed in a stable universe, told him, *"Vos calculs sont corrects, mais votre physique est abominable."* ("Your calculations are correct, but your physics is atrocious.") British astronomer Fred Hoyle, who subscribed to Steady State theory, came up with the phrase "big bang" to mock Lemaître's idea.

The church liked the idea because it implied a biblical creation event, and some scientists were suspicious that a priest might have invented it purely for religious reasons. Lemaître did think that God had created the place at which the Big Bang occurred. He saw it as a huge single atom, twice the distance from here to the Sun in diameter, which he called the Primeval Atom. Modern scientists perceive it as a singularity, a dot with no mass but plenty of energy. The discovery of cosmic background radiation, which Lemaître lived long enough to see, supported his Big Bang Theory rather than Hoyle's Steady State.

Lemaître's paper was eventually translated and distributed more widely by his old teacher in England, Arthur Eddington, and Big Bang Theory is accepted widely by the scientific community and the wider public. Lemaître was nominated for a Nobel Prize in 1954 for his version of the expanding universe, and again in 1956 for his vision of a Primeval Atom. As for who or what made that, or the singularity, well that's another question.

# Alexander Fleming

## (1881–1955)

## Penicillin

Penicillin has transformed medical treatment and saved millions of lives. It's surprising, therefore, that no one wanted to know about it when Alexander Fleming first discovered it; and that he was far from the first to consider its potential as a healing agent.

Fleming, a Scottish microbiologist working in St Mary's Hospital in London, discovered penicillin by accident in 1928, when a staphylococcal bacterial culture he had been growing was contaminated by a spore of the fungus *Penicillium rubens*. Where the fungus, a kind of mould, had established a colony, the growth of the bacteria was inhibited. Mould was killing the bacteria.

By 1928, it was well established that bacteria caused disease. The German microbiologist Robert Koch was the first to prove it, by identifying that *Bacillus anthracis* was the cause of anthrax, in 1876. The following year French biologist Louis Pasteur noted that the anthrax bacillus was inhibited by an unidentified mould. The observation was repeated by several other European scientists in the later years of the century, and in 1920 two Belgians had even identified *Penicillium* as the mould. Their paper was largely ignored.

These relatively modern scientists were in turn merely recording in a more methodical way what physicians from earlier times had already known, even if they didn't understand the process. In ancient India and Egypt medical men and women regularly used flora and fungi to heal infections. In seventeenth-century Poland, wounds were treated with a poultice of wet bread and spiders' webs. Even in late-nineteenth-century France, Arab stable hands applied mould to horses' sores.

Fleming's claim to fame is justified because he was able to repeat the contamination experimentally and to understand what was happening microbiologically. He saw the potential in it, and he tried this new antibacterial agent (which he named penicillin) on several different bacteria. Besides staphylococcus it worked against streptococcal species and the bacterium that causes diphtheria; but it was ineffective against typhoid and flu.

Fleming asked some of his colleagues to isolate the active chemical compound in his penicillin, which at that stage was no more than fungus soup. When they couldn't, he lost interest in further research into the phenomenon. But a student of his, Cecil Paine, became the first man to use penicillin as a treatment when he cured an infant of an eye infection, ophthalmia neonatorum, in 1930.

Pure penicillin was finally isolated in 1940, by a research project headed by Ernst Chain and Howard Florey, and two years later Alexander Fleming used it for the first time, in the treatment of meningitis. Early mass production of the drug was driven by the need to keep healthy the thousands of Allied troops fighting on the front lines of World War II.

Penicillin has been a victim of its own success. It is so effective, and so widely prescribed, that some bacteria have developed a resistance to it. Nevertheless it has been described as "the single greatest victory ever achieved over disease". Fleming, who with Chain and Florey shared the 1945 Nobel Prize in medicine, was typically self-effacing about his achievement: "When I woke up just after dawn on September 28, 1928," he recalled, "I certainly didn't plan to revolutionize all medicine by discovering the world's first antibiotic, or bacteria killer. But I suppose that was exactly what I did."

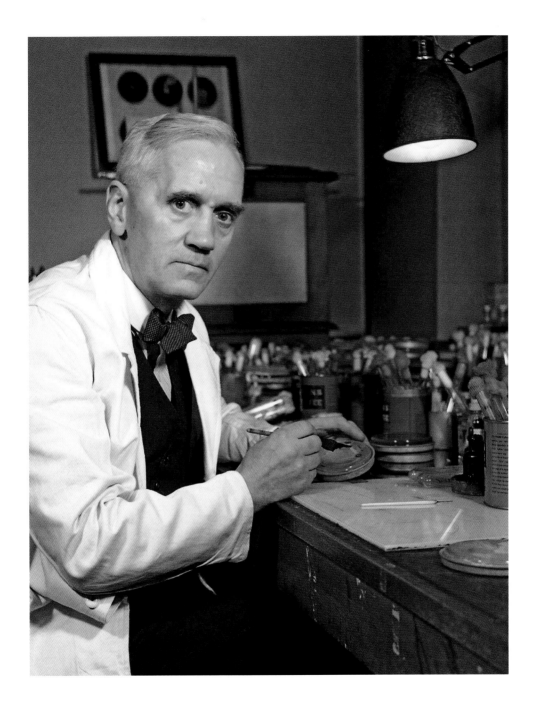

*ABOVE: Sir Alexander Fleming photographed at his workbench in the early 1940s.*
*OPPOSITE: Genetically modified fungi on an agar plate, part of the antibiotic production process.*

ABOVE: *An argon laser used in Raman scattering experiments.*

LEFT: *Sir Chandrasekhara Venkata Raman in the late 1960s. Raman had been encouraged by the findings of American physicist Arthur Compton, who found that X-rays could be described as particles. He reasoned that there would have to be an optical analogue to the Compton Effect.*

# C. V. Raman

## (1888–1970)

## Light imparts energy to molecules

At the age of sixteen, Chandrasekhara Venkata Raman's precocious talent caused a Nobel Prize winner to mistake him for a professor. The young Indian physicist became a professor at the age of twenty-nine, shed light on the colour of the sea and gave the world a non-invasive diagnostic tool for conditions including cancer.

C. V. Raman was the top of his class at every stage of his education and was so sure in 1930 that he would win a Nobel Prize that he booked his passage from India to Sweden four months in advance of the announcement of the recipients. His first two articles, published in the *Philosophical Magazine* while he was still an undergraduate at the University of Madras, prompted the 1904 Nobel Prize winner Lord Rayleigh to begin a correspondence with him. It was Rayleigh who addressed him, either in error or in amused respect, as "professor".

Raman and Rayleigh shared an interest in light and sound waves. Rayleigh was at the time investigating the sensitivity of the human ear to the direction from which sounds come, while Raman's early work was in the vibrations in drums and stringed instruments. Rayleigh had earlier in his career correctly explained why the sky is blue – it's because of an effect now called Rayleigh scattering.

He incorrectly explained that the sea is blue because it reflects the sky; and as Raman travelled to and from England in 1921, on a ship passing through the Mediterranean Sea, he began to question Rayleigh's explanation. When looking at the water's surface through a prism that eliminated reflected light, the sea actually looked more blue than ever, and Raman speculated that refraction from particles within the water, rather than reflection from the surface, might be the reason.

This proved to be the case and Raman began to investigate why the refraction would present the colour blue rather than any of the others of the rainbow. At first

he suspected some form of fluorescence, although this was unlikely because the refracted light was polarized. His breakthrough came after Arthur Compton in the US discovered that electromagnetic waves (of which light is a form) could be considered as particles. Compton was working with X-rays; but Raman saw at once that it must be true of light also.

He and his research assistant K. S. Krishnan began to experiment at the start of 1928, shining light of just one colour of the spectrum onto transparent liquid. They discovered, using a spectrograph of Raman's invention to detect electromagnetic waves, that when light of one colour struck the liquid, atoms in the liquid emitted the same colour *and a second colour* lower in the spectrum than the original one as scattered light.

This became known as Raman scattering and was caused by the particles of light exciting the atoms, energizing them to this effect. It later emerged that the colour change can act as an identifier of particular molecules. Raman spectroscopy is widely used in research laboratories. It can be applied harmlessly to living cells and helps with the detection of cancers.

Not only had C. V. Raman discovered a new phenomenon, it was, as American physicist Robert W. Wood declared when recreating and confirming Raman's experiments, a "very beautiful discovery which … is one of the most convincing proofs of the quantum theory". It was one of the first results to confirm the quantum theory of light, and – as Raman always knew it would – won him the Nobel Prize for physics in 1930.

# Subrahmanyan Chandrasekhar
## (1910–1995)

## Massive stars can collapse under their own gravity

Can you tell your red giant from your brown dwarf? Which comes first, a black hole or a supernova? The universe is filled with stars at various stages in their life cycles and astrophysicist Subrahmanyan Chandrasekhar devoted his life to the study of the physical changes through which a star passes.

Chandrasekhar was educated by his parents at the family home in Lahore – then in British India, now in Pakistan. At the University of Madras he studied quantum physics and won a grant and a place at Cambridge University to pursue his postgraduate doctorate.

Cambridge astrophysicist R. H. Fowler was to be Chandrasekhar's supervisor at the university. During the ocean voyage to Britain to take up his place in 1930, Chandrasekhar passed the time by reviewing and correcting a leading work on the mechanics of electron gases in white dwarf stars, written by Fowler. Chandrasekhar made new calculations in the light of Einstein's special relativity.

All stars form from stellar nebulae, clouds of dust and gas from which protostars coalesce and grow. Stars shine because of the burning hydrogen of which they are largely composed. When they run out of hydrogen they start to fuse helium, expanding in size to become red giants or red supergiants depending on their mass. When red giants, from smaller stars no bigger than our own sun, run out of helium they try to combine carbon but cannot. The gases spread out into space while a core of carbon collapses to become a white dwarf, which eventually runs out of energy and stops shining altogether, a brown dwarf.

Larger stars, much larger than the Sun, become red supergiants when they exhaust their hydrogen. Supergiants have enough energy to create elements and are eventually composed entirely of iron. At that stage the star collapses under its own gravity and explodes, briefly becoming a supernova, before coalescing as either a neutron star or – in the case of very large supergiants – a black hole.

Black holes had yet to be discovered in 1930 and many astrophysicists believed them to be impossible. But Chandrasekhar's results while sailing to England suggested that there must be an upper limit to the size of white dwarfs. The Chandrasekhar Limit defines the maximum mass of a white dwarf (currently estimated at 1.44 times the mass of our Sun). Anything larger will not be stable as a white dwarf but continue to collapse and become a supernova. This in turn implies the difference between a red giant and a red supergiant, and illuminates our understanding of supernovas, neutron stars and – now that they have been discovered – black holes.

Not everyone accepted Chandrasekhar's idea of a limit. The leading English astronomer of the day, Arthur Eddington, publicly dismissed Chandrasekhar's Limit, an act which Chandrasekhar believed was at least in part racially motivated. Eddington was an influential figure and although others saw the sense of Chandrasekhar's theory, few spoke up. His Limit was ignored for many years until mainstream astrophysics caught up with him, and he was at last awarded a Nobel Prize for physics in 1983. In 1999 his widow donated a sum equal to his Nobel Prize money to the University of Chicago to establish a Fellowship in his memory.

*ABOVE: Planetary nebulae surrounding a white dwarf.*

ABOVE AND LEFT: *Two images of the charismatic chemist.*

OPPOSITE: *Pauling's book on the nature of chemical bonds has remained in print since 1939. He purposely left out almost all mathematics and concentrated on description and real-world examples. The book was filled with drawings and diagrams of molecules and was very readable for a science textbook. His intention was that chemistry should be understood not memorized.*

# Linus Pauling

## (1901–1994)

## Valence bond theory

It could be said that Linus Pauling and quantum mechanics grew up together. The "new physics" of quantum theory was taking shape during Pauling's student years; his was the quantum generation and he was one of its brightest stars.

Pauling came to chemistry early. One of his boyhood friends had a child's chemistry set, which fascinated the young Linus. At Oregon State University he studied chemistry – of course – and while still a student he was offered a job teaching a course in quantitative analysis, which he had only just finished.

In his final student years he was intrigued by the work of Gilbert Lewis who had recently published a new theory about the covalent bonds between atoms in molecules – the mechanism and rules by which atoms are able to join with other atoms. After receiving his doctorate, Pauling spent two years in Europe visiting centres of the new study of quantum physics, considering its impact on chemistry.

He took up a post at Caltech in 1927 and devoted himself to the study of complex molecules and their structure. Over the next five years he produced a career-defining body of innovative research, with over fifty papers to his name. He built on Lewis's work and combined it with the efforts by Walter Heitler and Fritz London to adapt Lewis to quantum mechanical theory. Then Pauling weighed in with his own ideas on valence bonding, namely resonance and orbital hybridization.

Pauling's *The Nature of the Chemical Bond*, published in 1931, pulled all these threads together to weave the founding description of modern valence bond theory. The expanded paper was published in book form in 1939 and immediately became the standard work. By 1969 it had been cited more than 16,000 times in scientific papers and it continues to inspire and influence current chemistry research today.

Pauling was not alone in his field and in his wake another concept emerged. Where Pauling's valence bond theory (VBT) proposed atomic orbital electrons which bonded with those of other atoms to form molecules, molecular orbital theory (MOT) posited orbitals which applied to the whole molecule. From the 1960s onwards MOT was incorporated in chemistry computer programmes and VBT lost favour. However, by the 1980s the problems of computing with VBT were being solved and it has since seen a revival.

Having laid the foundations of quantum chemistry, Pauling turned his attention to study organic molecules. His research in the structure of haemoglobin, proteins, amino acids and peptides inspired the British team of Watson, Crick, Wilkins and Franklin to discover the structure of DNA. His study of sickle cell disease marked the beginning of molecular genetics theory, and in the final decades of his remarkable career he investigated the structure of the atomic nucleus.

During World War II, Pauling declined Robert Oppenheimer's invitation to join the Manhattan Project. He fell out of favour with the scientific and political establishment for his very public repudiation of nuclear warfare and his campaigning against the Vietnam War. Pauling was awarded the 1962 Nobel Peace Prize, his second Nobel after the 1954 prize for chemistry. He remains the only person to have won two unshared Nobel prizes and one of only four to win two in different disciplines.

NEW! A VALUABLE TEXT *and* REFERENCE BOOK for CHEMISTS and MINERALOGISTS

The Nature of the Chemical Bond

*By* LINUS PAULING

429 Pages 72 Illustrations

*In* THIS BOOK

The author discusses the problems of chemical binding and the structure of molecules and crystalline aggregates of atoms and molecules from the modern point of view.

PRICE $4.50

PUBLISHED BY CORNELL UNIVERSITY . . . 124 ROBERTS PLACE PRESS ITHACA, N. Y.

# James Chadwick

## (1891–1974)

## The neutron

James Chadwick lived a full life: episodes include his accidental enrolment in physics, his training with two giants of nuclear physics, his time in a prisoner of war camp, his role in the Manhattan Project. He also discovered a new elementary particle.

Chadwick was a bright boy who shone at school and passed the entry exams for two universities. He intended to study mathematics but was too shy to correct the interviewing member of staff who thought he meant to learn physics. Thus, accidentally, began an illustrious career. Chadwick studied under the physicist Ernest Rutherford, who won a Nobel Prize in 1908, the year Chadwick enrolled.

After his graduation he won a scholarship to study in Europe and chose to work on beta radiation under Hans Geiger, head of radiation research in Berlin and the inventor of the Geiger counter, which detected radioactive particles. Unfortunately Chadwick was still in Berlin when war broke out in 1914 and he spent the next four years in a civilian internment camp there.

On his return he rejoined Rutherford in Manchester, and followed him when Rutherford was appointed director of the Cavendish Laboratory in Cambridge in 1919. Chadwick was eventually appointed Rutherford's assistant director and the two men worked closely on matters atomic.

Their special interest was the composition of the nucleus of the atom. It was believed at the time to consist of electrons and protons, a theory which gave the correct electrical charge and mass but the wrong "spin", a form of momentum found in nuclei and elementary particles. Rutherford and Chadwick theorized that instead of electrons there must be a new elementary particle with no electrical charge, which they called the neutron.

Research elsewhere set Chadwick on the road to proving the existence of the neutron. The husband-and-wife team of Frédéric and Irène Joliot-Curie believed in 1932 that they had dislodged protons from paraffin wax using gamma rays from beryllium and polonium. The Cambridge pair thought that gamma rays were not strong enough to have shifted the protons. Neutrons on the other hand would need very little energy to do the job.

As director, Rutherford had other responsibilities at Cavendish; but Chadwick dropped everything to recreate the Joliot-Curie experiment. The protons ejected from the wax behaved precisely as if they had been hit with neutral particles the size of protons. He had found the neutron, and the Joliot-Curies had not noticed it.

Chadwick moved fast to claim priority for the discovery, writing a letter only two weeks later headed "Possible Existence of a Neutron" and followed it three months later with an article, the headline shortened to "Existence of a Neutron". The discovery of the neutron single-handedly redirected the course of scientific research. Neutrons could bombard nuclei, where beta decay would convert them to protons. This made it possible to create new elements in the laboratory.

It also made it possible to split the atom, whether in nuclear power plants or bombs. Chadwick was asked by the British government to consider the possibility of a bomb at the start of World War II. His report was shared with the US, which promptly gave the go-ahead for the Manhattan Project to build its own bomb. Chadwick was present when Britain and America agreed on the need to bomb Japan, and witnessed the first test on 16 July 1945.

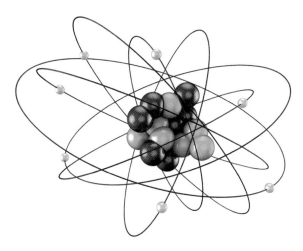

ABOVE: *The mushroom cloud following the explosion of the Trinity plutonium bomb on 16 July 1945 in Alamogordo, New Mexico. Chadwick was present to witness the explosive power of the atom.*

LEFT: *A nitrogen atom showing seven electrons orbiting seven protons and seven neutrons.*

*ABOVE: The Cockcroft-Walton generator. No longer in use, this is the device that was used to provide the initial acceleration to protons prior to injection into the 200 MeV linear accelerator (LINAC), for further acceleration before being delivered to the Alternating Gradient Synchrotron.*
*OPPOSITE: Ernest Walton (left) with Ernest Rutherford (centre) and John Cockcroft (right) in 1932.*

# Ernest Walton & John Cockcroft
## (1903–1995 / 1897–1967)

# Nuclear fission by proton bombardment

When two PhD students of Ernest Rutherford both decided to develop their supervisor's earlier experiments in splitting the atom, their efforts were unsuccessful until Rutherford suggested they work together.

Both Ernest Walton and John Cockcroft had shown such early promise in their physics careers that Rutherford agreed to supervise their PhD theses personally in the Cavendish Laboratory of Cambridge University. The students were naturally in awe of him: Rutherford had in 1919 become the first man to convert one element into another when he bombarded nitrogen with alpha particles from radioactive elements and turned it into oxygen. The event, remarkable in itself, also gave insights into the composition of atoms.

The use of alpha particles, however, only worked with a few specific elements. Walton hoped to achieve more consistent results by firing charged particles from a homemade particle accelerator into light nuclei such as lithium. Cockcroft, meanwhile, had built himself a discharge tube from which to accelerate protons into the nuclei. Rutherford advised them to join forces and found them a little university money – £1000 – with which to build a better machine.

Even in 1930 that substantial sum was not enough for a groundbreaking piece of scientific equipment, and the particle accelerator which Walton and Cockcroft now assembled was cobbled together from spare bicycle parts, lumps of clay, old food tins, glass tubes and anything else they could find in the laboratory. When it was finally replaced with a US-built cyclotron in 1938, the cost including a building to house it in was over £250,000.

Nevertheless their original design, which became known as the Cockcroft-Walton Accelerator, did what it was required to. It was the world's first particle accelerator and it could produce 700,000 volts DC from a low-voltage AC current – more than enough to accelerate protons.

On 14 April 1932 Walton fired the protons into a thin sheet of lithium. As he recalled fifty years later, "I saw tiny flashes of light looking just like the scintillations produced by alpha particles which I had read about in books but which I had never previously seen." These alpha particles were the nuclei of helium atoms into which lithium atoms had been transformed. The Cockcroft-Walton accelerator had performed the first artificial nuclear fission in history.

Walton and Cockcroft carried out further experiments with a variety of particles including alpha particles and deuterons, splitting the atoms of many other elements including carbon and nitrogen and creating carbon-11 and nitrogen-13, radioactive isotopes of those elements. Their work confirmed many ideas about atomic structure which until then had only been theories; and it proved Einstein's famous mass-energy equivalence equation, $E=mc^2$.

John Cockcroft went on to play an important part in the development of the British nuclear power industry. Ernest Walton was appointed Erasmus Smith's Professor of Natural and Experimental Philosophy at Trinity College, Dublin in his native Ireland, where he had a reputation as a lecturer who could convey complicated material in an accessible, comprehensible way. The two colleagues shared the 1951 Nobel Prize in physics.

# Fritz Zwicky

## (1898–1974)

## Dark matter

Dark matter and energy make up around 95% of the universe. In other words most of the universe is invisible to us, even with the most powerful telescopes in the astronomical world. How do we even know it's there? Swiss astronomer Fritz Zwicky worked it out.

Dark matter, by definition, is hard to see in a dark universe; but often in modern astronomy it is possible to know things are present by their effect on other objects, for example their gravitational influence. Whatever dark matter is made of, it doesn't seem to have any so-far traceable influence of its surroundings *except* through gravity.

It was dark matter's gravity which convinced Fritz Zwicky of its existence. In 1933 he was studying the Coma Cluster of galaxies. He applied a calculation called virial theorem to the system, which compares its potential energy (the result of gravitational stresses between objects in the system) to its kinetic energy (the amount of movement actually taking place in the orbit and rotation of objects within the system).

Zwicky observed that the gravity required for the cluster's kinetic energy was more than 400 times greater than that which was apparently available from the visible elements of the cluster. This was evidence that something else, massive and unseen, was exerting a gravitational influence within Coma. Zwicky called it "dark matter".

The idea of unseen matter in the universe was first mooted in 1884 by William Thomson, Lord Kelvin, who suggested of our own galaxy that "many of our stars may be dark bodies". Only a year before Zwicky's discovery,

Dutch astronomer Jan Oort saw a gravitational anomaly between the rotation of stars and the mass of visible objects within the group of galaxies of which ours is a member; but it was dismissed as an error of calculation.

In fact Zwicky's own calculations were based on incomplete data; but his logic was sound. Over the rest of the decade similar observations were made by other astronomers including Jan Oort (again) who saw a similar anomaly in the Spindle Galaxy (where in 1992 a supermassive black hole was also discovered). Further evidence of unseen gravitational forces came to light, as it were, in the 1970s and 1980s, including the gravitational bending of light emitted by objects behind such galaxies.

Needless to say, the existence of dark matter challenged existing models of the universe and its formation. Estimates vary, but it is currently believed that dark matter forms 30% of the total mass of the universe. Dark energy and matter together make up between 90 and 99% of it.

And yet it has never been observed directly, only inferred. The consensus is that dark matter consists of a so-far undiscovered sub-atomic particle, comparable but completely different to those which are at the heart of every nucleus of every atom in the visible universe. The search for this particle is at the heart of contemporary particle physics.

*ABOVE: A majestic face-on spiral galaxy located deep within the Coma Cluster of galaxies, which lies 320 million light years away in the northern constellation Coma Berenices. The galaxy, known as NGC 4911, contains rich lanes of dust and gas near its centre.*

*OPPOSITE: The feisty Fritz Zwicky. A fellow astronomer at Caltech, Walter Baade, named a galaxy after himself that was discovered by Zwicky. Edwin Hubble corrected the record and the galaxy was catalogued as a Zwicky galaxy.*

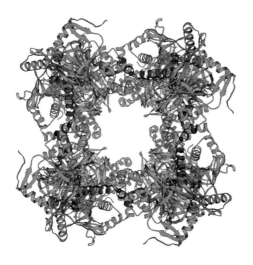

*ABOVE: Hans Krebs is best known for his discoveries of two important chemical reactions in the body, namely the urea cycle and the citric acid cycle (AKA the Krebs Cycle). With Hans Kornberg, he also discovered the glyoxylate cycle.*

*LEFT: A Molecular model of the enzyme dihydrolipoamide succinyltransferase, which is involved in the citric acid (or Krebs) cycle*

# Hans Krebs

(1900–1981)

## The Krebs Cycle

The metabolic cycle is the sequences of processes by which the body converts nourishment into energy and disperses waste by-products. Hans Krebs devoted his career to unravelling the complex interdependency of bodily functions on which we unconsciously rely.

Krebs fought briefly in World War I before going to university, an experience which he said taught him the value of disciplined record keeping and teamwork. After he graduated in chemistry he developed an interest in biochemistry and worked for a while for Otto Warburg, one of the great biochemists of the twentieth century. Warburg was an old-fashioned German, autocratic, formal, intolerant of errors – he served in the cavalry during the war and won an Iron Cross for bravery. But he had learned the same lessons as Krebs about discipline, focus and integrity; Krebs greatly admired him and learned a great deal from him in terms of methods and rigorous standards of research.

Krebs moved to the University of Freiburg where he had the freedom to pursue his own research. He refined a Warburg technique which used thin slices of tissue to study cell metabolism. Krebs added a blood plasma substitute of his own invention. This improved the accuracy of experiments and is still used all over the world today.

With his Freiburg research student Kurt Henseleit, Krebs uncovered the metabolic cycle that converts ammonia – a by-product of the breakdown of amino acids – into urea, making it possible to excrete unwanted nitrogen from the body with urine. It was the first metabolic cycle to be described in detail, and in 1932 it earned Krebs immediate international renown.

His fame in Germany was short lived. As soon as Hitler came to power in Germany in 1933 he passed a law making it illegal to employ Jews in professional occupations. Krebs was sacked in April. Fortunately he had an admirer in Sir Frederick Hopkins of Cambridge University who, learning of Krebs' situation, offered him a post in Cambridge. Unexpectedly Krebs was allowed to take his experimental equipment and samples with him and by July 1933 he was continuing his research in Cambridge's department of chemistry. Two years later he transferred to the University of Sheffield, where he did his work of greatest biochemical importance.

There Krebs uncovered the enormously important tricarboxylic acid cycle, also known as the citric acid cycle or the Krebs cycle. The eight-stage cycle starts with the breaking down of acetyl co-enzyme A by oxaloacetate, and finishes (after producing water, carbon dioxide and energy) by regenerating oxaloacetate for the next turn of the cycle. This cycle processes about two thirds of all the food we eat; and it connects with almost all other metabolic cycles.

The importance of identifying the processes of the Krebs cycle cannot be overstated. It is absolutely central to the functioning of the human body and of many other animals. Hans Krebs shared the 1953 Nobel Prize for physiology with Fritz Lipmann, a fellow German Jew who had emigrated to the US and was the discoverer of the co-enzyme A involved in the Krebs cycle.

# B. F. Skinner

## (1904–1990)

## Operant behaviourism

The Skinner Box, a way of rewarding desired behaviours, is a standard tool of investigations in animal psychology and perception. The experiments of Burrhus Frederic Skinner, although mostly conducted with animals as the subjects, led to discoveries about human behaviour too.

During an unsuccessful teenage year in which he attempted to follow his youthful dream and become a writer, B. F. Skinner stumbled across the works of Russian physiologist Ivan Pavlov and American psychologist John B. Watson, founder of the Behaviourism school of psychological thought. Behaviourists believe that we learn to behave in a certain way as a result of historical reinforcement of that behaviour. In short, if children are praised for washing their hands, they will continue to wash their hands.

Skinner abandoned his work on the great American novel and went to Harvard where, after earning his psychology PhD, he remained as a tutor. He was interested in psychological research conducted in a precise, scientific way. To this end he devised the Skinner Box, which he called an "operant behaviour apparatus" – a container which dispensed food or some other reward when an animal subject pressed a lever or pecked at a button.

To accompany the Box he invented a machine that recorded how often the reward was triggered by an animal's behaviour. By experimenting with different rewards, Skinner was able to observe a behaviour that was not a response to some stimulating event such as the bell that caused Pavlov's famous dogs (he claimed) to salivate; it was instead dependent on the reward which followed it.

Skinner called this operant behaviour, compared to respondent or Pavlovian behaviour. Respondent behaviour is automatic, whether it is in reaction to the sound of a bell or other such stimulus. Operant behaviour is not stimulated, but seeks to repeat a reward by a learned action.

He described these rewards as operant reinforcements, and they could be either positive (for example giving food or praise) or negative (for example removing a restraint or ending an unpleasant task). Punishment is another technique for operant conditioning, and again it can be positive (for example imposing imprisonment or physical punishment) or negative (for example removing a privilege or a favourite toy). Reinforcement strengthens the behaviour that it rewards; punishment discourages it.

Skinner understood operant behaviour as proof that free will was just an illusion. We are, he thought, all just products of the reinforcements in our earlier lives, and he believed that society could be engineered through the judicious use of appropriate rewards. He presented his fullest vision of this psychological utopia in fictional form, in *Walden Two*, a novel published in 1948 which described citizens conditioned by operant reinforcement rejecting the need for free will. The title is a reference to Thoreau's *Walden*, which pictures a utopia closer to nature than Skinner's.

*Walden Two* inspired the establishment of more than a dozen real-life communities. Behaviourism is no longer the dominant psychology it once was; Skinner's operant behaviour however remains pertinent and is still used in many situations from mental health work to animal training.

*ABOVE: Pigeons were often used in Skinner experiments. He taught them to play ping-pong and peck at levers and, during World War II, he was asked to investigate the possibility of a pigeon-guided missile.*

*RIGHT: Skinner wasn't interested in understanding the human mind. His field of study, known as behaviourism, was concerned with observable actions and how they arose from environmental factors.*

*ABOVE: Enrico Fermi checking the electrical circuit of a neutron counter in Chicago, 1948.*

*LEFT: Work on the Manhattan Project was not confined to Chicago. This photo from February 1945 was taken at the Hanford site on the Columbia River in Washington State. It shows the vertical safety rods and the cables that support them at the top of the atomic pile of a nuclear reactor. The silver-coloured drums in the background contained boron solution, part of an emergency shutdown system should the rods be blocked by an earthquake.*

# Enrico Fermi

## (1901–1954)

## Nuclear fission by neutron irradiation

Although often remembered as the architect of the atomic bomb, Enrico Fermi's contributions to the nuclear age are much more fundamental. It was for those that he received his Nobel Prize; and it was the Nobel Prize which probably saved his family from the anti-Semitic horrors of World War II.

Fermi is often portrayed as non-political, but he was shrewd enough to use the excuse of the Nobel award ceremony to get out of his native Italy at the point in 1938 where Benito Mussolini was adopting the anti-Semitic policies of his ally Adolf Hitler. Fermi's wife was Jewish, and he used the Nobel Prize money to emigrate with his family to the United States. Fermi also took a public stand against President Truman's decision to develop a nuclear bomb after World War II.

Fermi's early work was in subatomic particles. He discovered new statistics in 1927 about a group of them, now named fermions in his honour, which follow Wolfgang Pauli's exclusion principle. They include electrons and protons, and others – neutrons – which at that time were yet to be discovered.

Pauli had proposed the existence of a mysterious particle which, to satisfy the law of conservation of energy, must accompany the electron which leaves the nucleus of an atom during beta decay. Fermi named it the neutrino and, assuming that Pauli was correct, developed a theory of beta decay which proved it to be an example of the weak nuclear force. He was proved right when, two years after his death, the tiny neutrino was finally detected by American physicists Clyde Cowan and Frederick Reines.

Two events in the early 1930s had set Fermi on the path to his Nobel Prize: the English physicist James Chadwick discovered the neutron, one of the components of an atomic nucleus, in 1932. The following year Frédéric and Irène Joliot-Curie, the daughter and son-in-law of the great Marie Curie,

found that they could artificially create radioactivity with alpha particles, the nuclei of helium atoms.

With his understanding of fermions, Fermi believed that neutrons would be more effective than alpha particles in this respect. He bombarded over sixty different elements with neutrons and found that not only did the process produce many new radioactive isotopes; he observed that neutrons were even more effective when they were slowed down. For this work on neutron-induced nuclear reactions Fermi was awarded his Nobel Prize.

Fermi continued his nuclear research after his emigration to the US. He was enlisted in the Manhattan Project to develop America's atomic bomb. He discovered the possibility of a chain reaction, releasing great quantities of energy, if uranium neutrons were fired into uranium that was already undergoing fission. In the empty squash courts beneath Chicago's Stagg Field grandstands, he devised and built a nuclear reactor and, on 2 December 1942, he set off the world's first controlled nuclear chain reaction.

After Truman approved the development of a hydrogen nuclear bomb, Fermi continued to work on the project in the hope of proving that it would be impossible. He believed the use of any such bomb would amount to genocide. Late in life he mused on the benefits of the application of scientific discoveries in society, which he believed was generally a good thing. "What is less certain," he added, "and what we all fervently hope, is that man will soon grow sufficiently adult to make good use of the powers that he acquires over nature."

# Stanley Miller

## (1930–2007)

## The origins of life on Earth, perhaps

Since Charles Darwin first presented the principle of evolution, one question has remained. While physicists have honed their theories of the origin of the universe – what was biology's Big Bang? How did life on Earth begin? An intriguing set of experiments in the 1950s offered a possible answer.

There's a window of a little over a billion years between the probable date of the formation of the Earth and the oldest known fossils. What happened in between? Was life on Earth born on Earth? Or was it carried here on an asteroid? Did it arrive fully formed, or did the component parts evolve and combine over many millennia?

Stanley Miller graduated with a chemistry degree from Berkeley in 1951. He attended a lecture by Harold Urey, who won a Nobel Prize for the discovery of deuterium, about the origins of the Solar System and the atmospheric conditions on Earth which might have enabled the synthesis of organic compounds. Miller's curiosity was aroused and he asked Urey to supervise his PhD thesis on the subject.

Together they designed a laboratory experiment to recreate those lifeless early conditions with a soup of four basic ingredients – water, methane, ammonia and hydrogen – into which they injected lightning in the form of a high voltage electrical discharge, in the hope of triggering a chemical reaction. Over the course of a week Miller found no fewer than five amino acids in the soup, which had been formed from these four inorganic compounds. Miller also found that, under heat, amino acids combine to make protein chains. He had successfully created some of the building blocks of life.

Something was missing, however. Life has to be able to reproduce itself, and proteins can't. For that you need ribonucleic acid (RNA) or deoxyribonucleic acid (DNA), complex molecules which can make copies of themselves. By coincidence Miller's first experiments were published in the same year that Rosalind Franklin, Maurice Wilkins, James Watson and Francis Crick discovered the structure of DNA. Miller spent the rest of his life refining and developing his experiments, but

never managed to create life itself; and nor has anyone else. His original experiments, however, galvanized a branch of biological science which until then had been considered peripheral.

His work has not been universally accepted. Analysis of early geology suggests that there wasn't much methane or hydrogen on Earth at the time, although supporters point out that they might have been locally abundant, near volcanoes for example. Critics also point out that living organisms are not simply stacks of chemical compounds: those chemicals are contained in cells.

There is somewhere on Earth which offers conditions similar to those that Miller recreated. Undersea vents emit hot gases that combine with salt water to produce deposits rich in minerals and hydrogen. The resulting warm, alkaline environment facilitates chemical reactions which, like Miller's soup, produce complex organic compounds. A research team from University College London successfully recreated these conditions in 2019 and produced primitive fat cells. It may be that deep-sea locations and not shallow puddles were the original laboratories of life.

*RIGHT: Stanley Miller, American chemist creator of the famous "primordial soup" experiment that advanced the most convincing explanation of how life originated on Earth.*

*LEFT:* Rosalind Franklin was assigned to the King's College DNA research project because she was the most experienced experimental diffraction researcher in King's at the time. It is only in recent years that her contribution to the DNA discovery story has been fully credited.
*BELOW:* A DNA helix molecule against an abstract background.

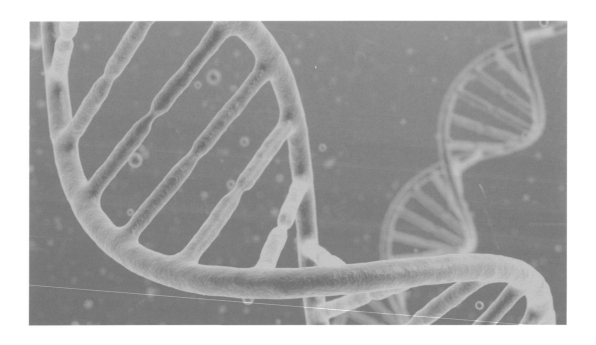

# Rosalind Franklin, Maurice Wilkins, James Watson & Francis Crick

(1920–1958 / 1916–2004 / born 1928 / 1916–2004)

## The spiral structure of DNA

It has long been recognized that physical characteristics like blue eyes and red hair are transferred from generation to generation. As mankind looks ever deeper into the cosmos in search of its origins, it has conducted a simultaneous search deep within for the biological mechanism for that transfer.

If genes are the building blocks of life, then DNA (deoxyribonucleic acid) is the instruction manual for assembling those blocks. Compared to proteins, the other components of our character-defining chromosomes, DNA is a relatively simple affair: it has four sub-groups compared to protein's twenty. The discovery of the structure of those groups was a vital first step in unlocking the genetic codes that they carry.

American biochemist Phoebus Levene did significant early work on DNA and RNA (ribonucleic acid) in the first decades of the twentieth century. British bacteriologist Frederick Griffith demonstrated the genetic role of DNA in a 1928 study of bacteria. And another Brit, molecular biologist William Astbury, used X-ray photography to reveal that DNA had a regular structure.

In the early 1950s two teams of British scientists were racing to map DNA. At Imperial College London, Raymond Gosling, a graduate student of Rosalind Franklin, took the clearest X-ray image yet in May 1952. It showed clearly a helical spiral structure.

Franklin discussed its implications with her colleague Maurice Wilkins and he showed it to Francis Crick and James Watson of Cambridge University, who were working in the same field. Drawing on data from Franklin and other researchers, Crick and Watson constructed a molecular model of rods and balls which conformed to all the known information about DNA. The result was in the form of two intertwined, interconnected spirals: an aesthetically pleasing and scientifically accurate double helix.

Crick is said to have burst into his local pub, the Eagle in Cambridge, at lunchtime on 28 February 1953 to announce with excitement: "I have found the secret of life!" Articles by Franklin, Wilkins, Gosling, Crick and Watson appeared side by side in the influential science journal *Nature* in March 1953, announcing the result of their research.

Further studies and experiments confirmed the validity of the Crick-Watson model and in 1962 Crick, Watson and Wilkins received the Nobel Prize for their discovery. Only living scientists are eligible for Nobel prizes, and Franklin had died in 1958; but it was regrettable that none of the awarded men cited her groundbreaking work in their Nobel lectures. Only Maurice Wilkins acknowledged her contribution in his acceptance speech, and the full extent of Franklin's contribution has only emerged subsequently.

Now that the structure of DNA had been confirmed, molecular biologists turned their attention to cracking the genetic code embodied in it. Once they had that, they could map the genome, first of fruit flies (which have only four pairs of chromosomes) and eventually of humans (who have twenty-three).

The discovery of the structure of DNA has led to insights into the development of species and of life itself. It offers understanding and potential cures for genetic disorders. DNA tests are now routinely applied to solving crimes. But with this new knowledge come ethical risks. Society continues to wrestle with the moral dilemmas posed by our ability to engineer species genetically.

ABOVE: *Fred Sanger in his
beloved greenhouse at home
in Swaffham Walbeck,
Cambridgeshire, in 1993.*
LEFT: *A vintage syringe for
injecting insulin.*
OPPOSITE: *An illustration
of DNA sequencing, for
which Fred Sanger created
the Sanger Method.*

# Frederick Sanger

## (1918–2013)

## The sequence of amino acids in insulin

One of only four people to receive two Nobel prizes, Frederick Sanger was, he claimed, "just a chap who messed about in a lab". When he was offered a knighthood for his remarkable biochemical achievements, he declined. "A knighthood makes you different, doesn't it," he said, "and I don't want to be different."

Sanger *was* different, and he is regarded as one of the finest biochemists of the twentieth century. His Nobel awards were for two of the most significant events in any scientific discipline in recent years – working out the structure of insulin, and becoming the first man to sequence a complete genome, a bacteriophage.

Type 1 diabetes was fatal in the nineteenth century. The link between insulin and diabetes was established in a series of discoveries. In 1869, Paul Langerhans, a pathology student in Berlin, first identified the clumps of cells in the pancreas which are now called the Islets of Langerhans. He didn't know what they were, but experiments by Eugene Opie in 1901 established that their destruction caused diabetes. Efforts by several research projects to identify the active secretion of the islets were hampered by World War I, although it was named insulin by the English endocrinologist Edward Sharpey-Shafer in 1916.

In the first years of the 1920s methods were devised for extracting insulin from cows, and in 1922 a dying diabetic patient recovered fully after treatment with insulin. Purified animal insulin thereafter became the standard treatment of the disease, and gradually scientists learned more of its nature: it was established in 1935 that insulin is a protein hormone containing the amino acids phenylalanine and proline.

Frederick Sanger, who spent his entire professional life in and around Cambridge in England, went to work on the amino acid structure of insulin in the early 1950s. He used partition chromatography and a new technique, paper chromatography, to identify that ox insulin consisted of two chains of amino acids and went on to describe the complete sequence of each in 1954, and the mechanism by which they are linked.

Type 1 diabetes is typically a disease of the young, developing in around 80,000 children and young adults every year. Sanger's work made the mass production of synthetic insulin possible, and has saved hundreds of thousands of lives. For his sequencing of insulin Sanger received the 1958 Nobel Prize for chemistry.

Subsequently, human insulin was genetically engineered for the first time in 1978, from a strain of the yeast bacterium *Escherichia coli*; and most insulin treatments are now of synthetic human form. In 2010 a team of Canadian scientists found a cheaper way to produce insulin from safflowers, a relative of the humble daisy.

After his breakthrough with insulin, Sanger went on to work in the sequencing of RNA and DNA for the British Medical Council. He devised a method for the fast, accurate sequencing of DNA molecules and used it to determine the complete DNA sequence of a bacteriophage, a virus that infects bacteria. That won him his second Nobel Prize in 1980, and he is one of only three double Nobel winners to earn both prizes in the same category. The Sanger Method was the one used to sequence the whole human genome in 2003.

# Yang Chen-Ning & Lee Tsung-Dao

## (born 1922 / born 1926)

## The non-conservation of parity in weak nuclear interactions of particles

*If there are two things that can be relied on in modern physics, they are the conservation of energy and the conservation of parity. Or so everyone thought. In the matter of parity, it turns out that some elementary particles in quantum mechanics can tell the difference between right and left.*

Scientists are able to observe elementary particles in particle accelerators such as the one at CERN, and in cosmic rays which break down as they strike the atmosphere. Some of these subatomic particles are elusive because they are unstable and quickly decay into other particles. Mesons, for example, start to decay after only a few hundredths of a microsecond.

There are several types of meson and two of them presented something of a paradox to particle physicists in the 1950s. Tau mesons and theta mesons seemed to be identical in mass and in decay time; they differed only in what they became after decay. Tau mesons each became two pions while thetas became three. By the rule of conservation of parity (a symmetry in the physical attributes of particles in what are called left-handed and right-handed systems), those different outcomes meant that tau and theta mesons must be different particles. But their identical masses and rates of decay meant that they must be the same.

Yang Chen-Ning and Lee Tsung-Dao, two Chinese physicists studying at Princeton on research fellowships, tried to reconcile this conflict, technically known as a violation of parity. The conservation of parity was at the time as fundamental to physics as the conservation of energy, and in particle physics this had been proved by experiment and observation.

But it hadn't. In an embarrassingly serious lack of experimental rigour, it had only been proved for certain particles, and the "rule" was assumed to apply to all. Particles interact thanks to one of four fundamental forces; and the conservation of parity had only been tested with those bound by electromagnetism and the so-called strong nuclear force, which are the dominant forces in most physical interactions. Mesons, however, use the weak nuclear force; and Yang and Lee found that no one had proved parity conservation for that. (The fourth force is gravity.)

To rectify this oversight, Yang and Lee devised a series of experiments which proved conclusively that parity need not be conserved in particles bound by the weak force. The failure to understand this earlier had held up particle research by several years. For their willingness to challenge the conventional wisdom, and their proof of non-conservation in weak decay, Yang and Lee shared the 1957 Nobel Prize for physics.

*ABOVE: Lee Tsung-Dao (left) and Yang Chen-Ning.*

# Jane Goodall

## (born 1934)

## Social structures in primates

Dr Jane Goodall, untrained but passionate about animal behaviour, changed the way we think about chimpanzees and all primates, including ourselves. Her methods were sometimes unconventional and controversial, but her lack of formal training left her with an open mind with which to observe.

Like many children, Jane Goodall had a dog and a cuddly toy, and liked to read. But she also had a pony, a tortoise and other pets; her cuddly toy was a chimpanzee, and her favourite books were Hugh Lofting's *Doctor Dolittle* stories, about a man who can communicate with animals. She also read Edgar Rice Burroughs' *Tarzan* series and from the age of eight she was fascinated by the so-called Dark Continent – Africa.

When at last she could afford to travel there at the age of twenty-three, she was put in touch with Louis Leakey, a British paleoanthropologist working in Kenya. Leakey was doing groundbreaking work in human genesis, and had already begun to establish that early man evolved in the African continent. Given the diversity of primates there, he was also interested in any clues they might give about human evolution.

He put Goodall to work in Tanzania's Gombe Stream National Park, a small reserve on the eastern shore of Lake Tanganyika, accessible only by boat. There she observed chimpanzees in the wild over many years, living the first fifteen of them almost permanently on the reserve. Lacking academic training and practice, she was able to observe her subjects uncompromised by conventions or preconceptions. She saw things that a qualified zoologist might not.

One of her most famous discoveries was that chimpanzees use tools; and not only use them but make or modify them. Goodall watched chimps insert blades of grass into termite mounds, pull them out covered in the insects and eat them. She also saw them breaking off twigs and stripping the bark from them to make them more effective. This overturned the conventional wisdom that only mankind was capable of making and using tools.

Despite her lack of undergraduate qualifications, Jane Goodall earned a PhD during the time she spent at Gombe. Although in the field she was admired for her rigorous record keeping and ethical behaviour, her supervisors at Cambridge University objected to her unconventional research methods and conclusions. Usually, animal subjects were known only by numbers, to avoid making any emotional connection with them, but Goodall named many of hers, including David Greybeard, the alpha male who first accepted her into his group. She remains the only human to have been granted this privilege.

Worse still, she identified character traits and exhibitions of emotion in the chimpanzees. Such "soft" observations were frowned on by a science – ethology – eager to establish its credentials as a "serious" study. "It was not permissible," she reflected later in life, "at least not in ethological circles, to talk about an animal's mind. Only humans had minds." But Goodall observed physical contact between chimps that was clearly comforting or playful.

Goodall corrected a long-held misapprehension that chimpanzees were vegetarian: they hunt other primate species for food. Moreover, she saw females not only killing the young of other females in the group but eating them. She also observed a behavioural trait once thought only found in man – the ability to wage war against another tribe. From 1974 to 1978 she saw two factions of a once-unified group of chimpanzees in a prolonged conflict, which ended only when all the males of one group had been killed.

*OPPOSITE: Jane Goodall observing chimpanzees in the Gombe Stream National Park, Tanzania.*

# Murray Gell-Mann & George Zweig

## (1929–2019 / born 1937)

## Quarks and gluons

Gone are the days when scientists believed that the atom was the smallest component of the universe. The average atom is 100 picometres in diameter; but peel away three further levels of sub-atomic particles and you find the quark – for the time being, one of the smallest things known to mankind.

In a sense, the whole history of science has been about discovering smaller and smaller particles that make up the fabric of our universe, and understanding their interactions. From the ancient view that all matter was essentially a variation of earth, air, fire or water to recognizing the first elements; from believing that atoms were the smallest possible things to discovering that they are composed of a nucleus and electrons.

Then we found that the nucleus itself is composed of neutrons and protons; and then that protons and neutrons are part of an array of sub-atomic particles which fall under the groupings of bosons, fermions or hadrons. (To complicate matters, neutrons and protons are part of a subgroup called baryons, which are both hadrons *and* fermions.)

Researchers at this miniscule level observed phenomena in the 1950s that suggested the existence of many other so-called elementary particles – pions, neutrinos, muons and more. These interact with each other in one of four ways – electromagnetism, gravitation, weak interaction (responsible for radioactive decay) and strong interaction (which binds most ordinary material together and provides weight by its sheer energy).

Murray Gell-Mann, an American physicist at Caltech, played a central role in understanding and classifying their behaviours. His study of recently discovered cosmic rays called kaons and hyperons led him to devise the concept of strangeness, a quality of particles interacting by electromagnetism or strong interaction which, Gell-Mann observed, slowed their radioactive decay. He originally noted that this behaviour was "strange", and the name stuck.

George Zweig was born in Moscow and emigrated to the United States. He studied particle physics at Caltech in the early 1960s, where his supervisor was not Gell-Mann but Richard Feynman. Zweig and Gell-Mann came separately to the conclusion that the properties of hadrons could be explained if they were considered to be composed of three even smaller particles. Zweig called the new particles aces; Gell-Mann named the new particles as quarks, after a word invented by James Joyce in his novel *Finnegan's Wake*. Gell-Mann's quirky name was the one that caught on.

Quarks, antiquarks, and gluons (the massless particles that bind them together) made sense of the properties of hadrons and were gradually accepted by the scientific community. The quark model for the structure of hadrons is now a fundamental basis of particle physics. Murray Gell-Mann was awarded a Nobel Prize in 1969 for his general contribution to the classification and interaction of particles; but Richard Feynman's nomination of Zweig and Gell-Mann for the 1977 Nobel for their discovery of quarks was unsuccessful.

LEFT: *Murray Gell-Mann.*

BELOW: *A record of the first observation of an omega-minus particle in 1964 (with an interpretive graph to the right). A negative kaon enters at bottom left and collides with a proton in the chamber's hydrogen to produce three particles, including the omega-minus, which consists of three strange quarks. The omega travels a short distance before decaying into a negative pion, which veers to the right across the bottom part of the picture, and a neutral xi particle which leaves no track.*

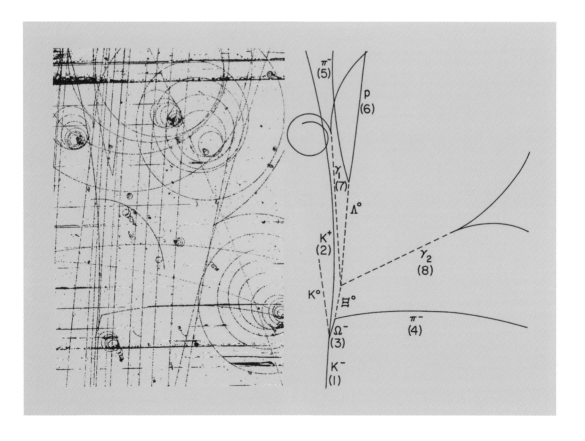

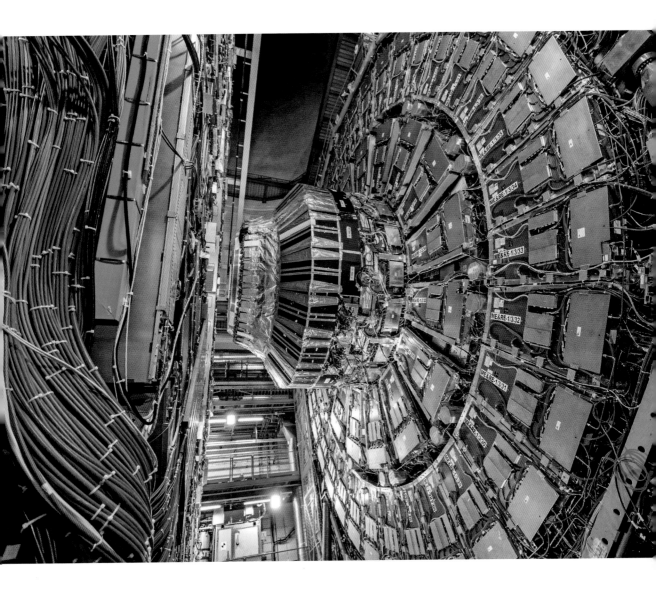

ABOVE: *The giant (14,000-tonne) Compact Muon Solenoid (CMS) detector in the Large Hadron Collider tunnel at CERN in Switzerland, where the Higgs boson was finally detected in 2012.*

OPPOSITE: *Peter Higgs, who along with other scientists in 1964 proposed the Higgs mechanism to explain why some particles have mass.*

# Peter Higgs
## (born 1929)
## The Higgs boson elementary particle

You don't need to know what a boson is. Quantum physics is the point at which even the most enthusiastic amateur scientist may be forgiven for leaving it to the professionals. You may, however, appreciate the remarkable insight that led to the discovery of the Higgs boson.

Bosons are a class of subatomic particle at the same level as fermions and hadrons. They are named after the Indian physicist Satyendra Nath Bose. With Albert Einstein, Bose devised the Bose-Einstein statistics, which define the characteristics of such particles. Their existence is a product of quantum theory.

Peter Higgs was born in the English city of Newcastle upon Tyne. His father was a sound engineer who undertook some of Higgs' early education at home. During World War II the boy attended a school in Bristol in the south-west of the country, of which a former pupil, Paul Dirac, was one of the founders of the field of quantum mechanics in the 1920s.

Higgs' own early interest was in molecular vibration, the subject of his 1954 PhD. That qualification earned him a post as Senior Research Fellow at Edinburgh University where, after a year, he became more interested in the altogether smaller particles of quantum physics. After a spell teaching in London he returned to Edinburgh as a lecturer and remained there for the rest of his career.

He became intrigued by the nature of mass. Since there was nothing before the Big Bang, particles can only have acquired mass after it. Higgs conceived of a process now known as the Higgs mechanism in which particles acquired mass. Quantum physics consists of a number of "fields", the quantum excitation of which generates fundamental particles of which the universe is ultimately composed; and Higgs

proposed not only a new field but also the new particle which resulted from its excitation – in scientific terms a "massive" (mass-giving) boson.

Other scientists had been converging on the Higgs Mechanism theory, although not on Higgs' proposed boson, and they began to incorporate it in subsequent theories. The theoretical existence of the Higgs field made sense of research results and even predicted the existence of other particles not yet found. When those particles, the W and Z bosons, were discovered, the search intensified for the elusive Higgs boson.

Because of the Higgs boson's role in giving mass to other particles in the fraction of a second following the creation of the universe, it was labelled the God particle in a popular science book published during the search. Higgs himself disapproved of the epithet; but as the authors of *The God Particle* argued in 1993, "this boson is so central to the state of physics today, so crucial to our final understanding of the structure of matter, yet so elusive" that its importance in the creation of our universe could not be overstated.

The Higgs mechanism and field remained working theories for decades until 2012, when the Higgs boson was at last identified in the Large Hadron Collider particle accelerator, 175m below the French-Swiss border. Its existence confirmed the existence of the Higgs field and the reality of the Higgs mechanism. Peter Higgs was present at the discovery. He was awarded the Nobel Prize for physics in 2013.

# Jocelyn Bell Burnell

## (born 1943)

## Neutron stars

Jocelyn Bell Burnell was the first person to observe a pulsar. Her astronomical achievement was overlooked because she was a woman and a student, before being recognized with science's richest award.

Jocelyn Bell was born in Lurgan, a town in Northern Ireland of strict traditional values. Boys and girls were separated for certain lessons at school because the school would not teach science to the girls. Luckily for Jocelyn her father was the architect who designed the nearby Armagh Planetarium. Visits to the planetarium had first engaged her interest in astrophysics; and her father's outraged protests to Lurgan College persuaded the school to change its no-science policy for Jocelyn and a few other ambitious girls.

Bell went on to study physics at Glasgow University and during postgraduate studies at Cambridge she helped to build the Interplanetary Scintillation Array (IPSA), a radio telescope designed to examine the recently discovered phenomenon of quasars. In 1967 she noticed an anomaly on the paper printouts from the IPSA, a series of strong regular pulses. Her supervisor Antony Hewish dismissed them as man-made errors, but Bell persisted and found other examples on earlier printouts.

She had discovered evidence of pulsars, neutron stars that spin at high speed on their axes. Eventually Hewish and his colleague Martin Ryle were convinced and all three wrote a scientific paper about the discovery. The regularity of the pulses is very useful to astronomers trying to understand and map the heavens, and pulsars may be responsible for very-high-energy cosmic rays.

In the ensuing publicity about the discovery, Bell was disgusted to observe sexism at first hand. Hewish answered the scientific questions and Jocelyn was asked about boyfriends. Worse was to come. Hewish and Ryle were awarded the 1974 Nobel Prize for physics – the first time the prize had been given for astrophysical work and Bell was completely omitted from the citation.

Bell was publicly forgiving about being overlooked – as a PhD student it was standard practice that her supervisor's name should have preceded hers on the paper – but she also noted Hewish's initial scepticism about the discovery, and meetings between Hewish and Ryle from which she was excluded. Her subsequent career was interrupted by marriage to Martin Burnell, and motherhood, but her true part in discovering pulsars is widely recognized by the scientific community.

Although she missed out on a Nobel, she has received many other honours. She won a Special Breakthrough Prize in Fundamental Physics in 2018, which came with a purse of $3 million, the world's richest science award. She has used the entire sum to set up the Bell Burnell Graduate Scholarship Fund to support the entry of more women, refugees and minorities into physics research.

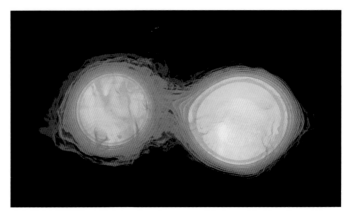

*ABOVE: Jocelyn Bell Burnell photographed in 1975 after the row over her omission from the Nobel Prize nomination.*

*LEFT: A pair of neutron stars colliding, merging, and forming a black hole. A neutron star is the compressed core left behind when a star explodes as a supernova.*

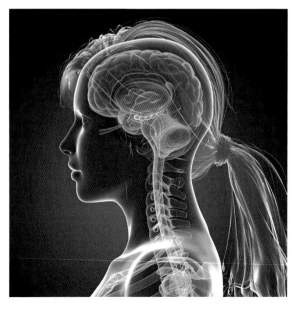

ABOVE: *John O'Keefe in the laboratory at University College London.*
LEFT: *The location of the hippocampus (marked in red) in the brain.*

# John O'Keefe

## (born 1939)

## Place cells in the brain

While scientists probe the furthest corners of the universe, the workings of the brain – like the seas and oceans – remain some of the least explored areas of our earthbound world. Neuroscience is slowly discovering how our minds perceive and interpret our surroundings.

Like a new continent, neuroscientists are mapping the brain, identifying regions, their functions and resources and the paths between them. They have discovered which area deals with language (the fornix), the words we speak (Broca's area), and the words we hear (Wernicke's area). They know that pain and temperature are processed through different routes than simple experiences of touch. There are different areas for different sensory memories – the auditory cortex for sounds, the visual cortex for images, and so on – which other areas then pull together to paint full pictures of memorable moments.

John O'Keefe, an Irish New Yorker, was awarded his doctorate in Montreal for a study of the way in which the amygdala receives sensory information. The amygdala is an area in the very centre of the brain which makes fight-or-flight decisions in moments of stress. O'Keefe moved to University College London to continue his research and has remained there ever since.

In London he transferred his attention to the hippocampus, an area near the amygdala at the top of the spine which decides which experiences to remember in the long term. It also coordinates emotional responses. O'Keefe was able to observe the responses of individual neurons (nerve cells in the brain) in the hippocampus of rats and correlate them with different behaviours.

He found cells that seemed to be associated with location. When a rat was in one place in a room, certain cells were active; and when it moved to another, different cells were activated. The rat's brain was mapping the room in neurons. Equally, damage to those cells in the hippocampus affected the rat's ability to know where it was.

O'Keefe and his student Jonathan Dostrovsky published their finding in 1971, that changes in behaviour resulting from damage to hippocampal cells was explained by the area's mapping function. He expanded on the theory in 1978 by affirming that the brain's spatial awareness was specifically rooted in the hippocampus. The idea met some resistance from other neuroscientists, although American psychologist Edward Tolman had first proposed the concept of cognitive mapping of knowledge, not just place, in his studies of rats in 1948.

The possibility of place cells in the brain gained traction after two Norwegian neuroscientists, Edvard and May-Britt Moser, spent two months in O'Keefe's laboratories in 1996. They had also been studying spatial awareness in relation to the hippocampus and back in Norway they began a long-distance collaboration with O'Keefe. They discovered cells near the hippocampus that were responsible for a sense of location – a sort of grid map in which to plot one's position and other important places. The Mosers also explored the mapping of position and relationship in a more social context, by studying the hippocampus in relation to social learning in rats.

The Mosers shared the 2014 Nobel Prize for physiology with John O'Keefe. The discovery of place cells has important implications for our perception of the world and offers vital insights into the condition of people who through illness or damage to the hippocampus get lost – sufferers of Alzheimer's disease, for example.

# Louise Webster & Paul Murdin

(1941–1990 / born 1942)

## Black holes

The theoretical possibility of black holes in space was first proposed in the eighteenth century. It was only sixty years ago, however, that astronomers Louise Webster and Paul Murdin could cautiously claim to have identified one.

An English church minister and amateur scientist, John Michell first proposed the idea of a black hole, which he described as a "dark star", in 1783. The name itself was coined in 1967; it was a reference to the Black Hole of Calcutta, a notorious prison from which no one ever emerged.

Michell, almost forgotten today, was a pioneering scientist who studied magnetism and devised a method for creating artificial magnets. He also suggested that earthquakes created seismic waves and were responsible for geological faults in the landscape, both of which we now know to be true. And he was among the first to argue that tsunamis were caused by undersea earthquakes.

In astronomy he was the first man to study binary stars (pairs of stars locked in orbit around each other); and in a 1783 paper he proposed the existence of stars with such strong gravity that even light could not escape from them – dark stars. Michell suggested that such a star would have to be 500 times larger than our Sun to have such a strong gravitational pull, something which Albert Einstein's General Theory of Relativity ruled out. They are now understood to be unimaginably compact: they are stars which have shrunk into extremely dense, heavy objects from whose gravity other objects cannot escape.

Michell's "dark stars" became more formally known as "gravitationally collapsed objects" and gradually scientists advanced and refined theories, which fitted what they knew about the phenomenon. Neutron stars are collapsed supergiant stars, and when Jocelyn Bell observed a pulsar (a kind of neutron star) in 1967, it accelerated interest in black holes.

Because they emit no light, black holes are by definition invisible to the naked eye. Englishman Paul Murdin and Australian Betty Louise Webster were working at the Royal Greenwich Observatory in England in 1971, looking for a strong but invisible X-ray source in the Cygnus constellation. It had first been observed during a space mission in 1964 and it proved to be a binary star. Because one of the pair was invisible, they speculated in the scientific paper on their research that it *might* be a black hole.

It was the first black hole ever discovered. The identification was confirmed in 1973, although famously Stephen Hawking bet a colleague in 1974 that it would prove not to be. He finally conceded defeat in 1990.

Black holes are important as significant stages in the life cycle of a star. They are evidence that the universe is constantly changing, and they have a huge impact on the surrounding material of the universe. The more a black hole swallows, the more powerful it becomes; and it is assumed that there are giant black holes, the weight of millions of stars, at the centre of many galaxies. The Messier 87 galaxy contains one such giant which became the first black hole to be photographed, in 2019, its presence revealed by the friction ring of heat around it caused by matter being sucked into its prison.

*ABOVE: An artist's impression of the Cygnus X-1 system. Cygnus X-1 is a stellar-mass black hole, a class of black holes that comes from the collapse of a massive star, with an immense gravitational pull.*

ABOVE: *The Arecibo Observatory radio telescope pictured before its untimely demise. Though it had withstood earthquakes and cyclones, in August 2020 one of the support cables snapped, sending the 900-ton instrument platform to its destruction below.*

LEFT: *Nobel-winning physicist Richard Feynman had taken a strong interest in the search for gravitational waves.*

# Russell Alan Hulse & Joseph Hooton Taylor Jr.

## (born 1950 / born 1941)

## Gravitational waves

Gravitational waves were recognized as a possibility by Albert Einstein's general theory of relativity. He thought they were too small ever to be observed. But they have been found, first by indirect implication, and then by direct measurement. They really are very, very small.

Disruptions to spacetime, such as the movement of large objects, the collisions of stars or the collapse of red supergiants, send ripples of gravitational waves (GWs) through the cosmos at the speed of light. By the time they reach the Earth they are very faint, like the light from distant stars; and they are a form of radiant energy similar to electromagnetic waves.

There was considerable debate about their existence. Even Einstein had doubts and wrote an article suggesting that they could not exist, although he later corrected himself. Scientists asked themselves whether gravitational waves, like ocean waves, carried energy and Richard Feynman proposed a thought experiment, the Feynman Gravitational Wave Detector: a rod with two beads on it which were able to slide freely along it with a little friction. As the waves arrived, they would move the beads further apart, generating heat from the energy in the wave and the work that the wave had done in moving the beads. You could in theory detect the wave by detecting the heat.

Science has subsequently invented more sensitive and accurate instruments. Joseph Weber, an American physicist, was the first to devise a serious GW detector, in 1969. But the frequency with which he reported detections cast doubt on its efficacy.

But at Puerto Rico's Arecibo Observatory in 1974 Joseph Taylor and his student Russell Hulse were conducting a survey of pulsars – rotating magnetic neutron stars whose electromagnetic radiation appears to pulse because of their rotation. They observed one pulsar whose pulse was regular but periodically stuttered slightly, coming either fractionally early or fractionally late.

It was a sign that it was a binary pulsar, one of two stars locked in orbit around each other, and with its neutron star pair it was the first ever to be found. Their subsequent analysis of it revealed an even bigger discovery: the binary's orbit was decaying, conforming to the loss of energy which would be associated with the emission of GWs. The Hulse-Taylor Binary Star was the first evidence, however indirect, of the existence of gravitational waves.

In the first decade of the twenty-first century a large-scale project, the Laser Interferometer Gravitational-Wave Observatory (LIGO), was set up to intensify the search for GWs. It first detected them in 2015, waves that were the result of the collision and merger of two giant black holes, together sixty-five times the mass of our sun. The movement detected by LIGO was minute, the equivalent of an arm 4 kilometres (2½ miles) long being tilted by one thousandth of the width of a single proton; or, to put it another way, the same as extending the distance from Earth to the nearest star outside the Solar System by the width of a human hair.

This collision of a binary black hole system was the first ever recorded. Since black holes form at the end of the life cycle of huge stars, it had immediate implications for the age of the universe. Now that they've been found, GWs are playing an important part in the exploration of space. They are evidence of their sources, be they binary stars, black holes or other big events – perhaps even the Big Bang itself.

# Benoit Mandelbrot

## (1924–2010)

## Fractal geometry and the Mandelbrot Set

Benoit Mandelbrot asked, "Can geometry deliver what the Greek root of its name seems to promise – truthful measurement, not only of cultivated fields along the Nile River but also of untamed Earth?" Geometry, he believed, was not just for the world's easy, manmade shapes, but also for its roughness.

"Clouds are not spheres," he observed. "Mountains are not cones, coastlines are not circles, and bark is not smooth, nor does lightning travel in a straight line." Mandelbrot, like most scientists through the ages, sought order in apparent chaos, and pattern in seemingly random behaviour. He saw the same jagged outline in maps of coasts as in a single rock on which he stood at the sea's edge; and he looked for ways to express that mathematically and geometrically. "The goal of science is starting with a mess, and explaining it with a simple formula," he said, "a kind of dream of science."

Often regarded as an eccentric, Mandelbrot attributed his unconventional perspective to his erratic mathematical education. In his early years he was taught in his native Poland by an uncle who rejected the unimaginative learning of rules and tables. Instead, he illustrated the magic of numbers through practical examples like maps and chess games, where maths could bring order to apparently unique sequences and settings.

After World War II he continued his studies in Paris and at Caltech and Princeton and eventually settled in the United States, working in the IBM laboratories in Yorktown, NY.

IBM gave him the freedom to explore his radical ideas and he remained there for 58 years. In that time he applied his convictions not only to nature but also to human activities, making

contributions in fields including meteorology, anatomy, linguistics, information technology, the social sciences and even the world of high finance. His book *The (Mis) Behavior of Markets* was reviewed as "the deepest and most realistic finance book ever published".

In the 1970s he returned to geometry via the world of Gaston Julia. Mandelbrot's uncle had introduced him to Julia's 1918 work in pure and applied mathematics; but as a headstrong young man Mandelbrot had rejected it. Now he saw its value in providing a set of equations, the Julia Set, which dealt with the reiteration of a rational function. With the benefit of IBM computers, Mandelbrot was able to develop new graphics programmes that used Julia to reveal exquisitely beautiful and evolving geometric forms.

Mandelbrot coined the word "fractal" for these forms in 1975, and in 1979 he devised his own equations, the Mandelbrot Set. Like the Julia Set, they produced complex, ragged-looking geometric shapes which exhibit the same intricate sequences at every scale, whether

viewed in tiny detail or as a whole. Romanesco broccoli and the leaves of ferns are two forms in nature that exhibit the same property.

Mandelbrot's 1982 book, *The Fractal Geometry of Nature*, introduced fractals into mainstream mathematics. His unique geometric perspective, born of maps and chessboards, has changed the way we see the supposedly chaotic worlds of nature and human behaviour.

*ABOVE: A computer-generated graphic using the Mandelbrot Set.*
*OPPOSITE: Looking like it was designed using Mandelbrot fractals, the beautiful*
*florets of a romanesco broccoli.*

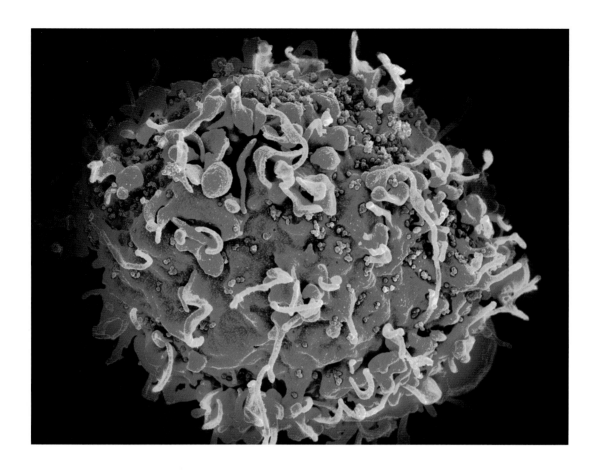

*ABOVE: A human T cell (green) shown under attack by HIV (yellow), the virus that causes AIDS. The virus specifically targets T cells, which play a critical role in the body's immune response.*
*LEFT: Polly Matzinger with Annie, who has contributed absolutely nothing to medical science, but can catch a frisbee.*

# Polly Matzinger

## (born 1947)

## The danger model in the immune system

The body protects itself from threats which would distress, damage or destroy it. The skin acts as a first line of defence; but inside there is an army of different cells working to detect, alert and counter-attack. This is the immune system.

The ancient Greeks noticed that if you survived one dose of a plague, you were immune from catching a second. In the late nineteenth century it was finally discovered that germs cause disease, whether airborne, waterborne or from contact. Louis Pasteur invented immunity through vaccination, and scientists began to study the immune system in greater detail.

At first it was thought that the body automatically rejected anything that wasn't part of it – the Self/Non-Self Model. American immunologist Charles Janeway proposed a more subtle version of that idea in 1989: his Infectious Non-Self Model claimed that evolution has left us with cells capable of distinguishing between infectious and non-infectious threats. It was an improvement, but it did not explain events such as the rejection of skin grafts, or infections by non-cytopathic viruses – viruses that don't damage cells. It fell to French immunologist Polly Matzinger to come up with a model which more successfully applied to all eventualities.

Dr Matzinger moved to the US with her parents when she was seven years old. She felt isolated at school and was voted Person Least Likely to Succeed by her classmates. It took her eleven years to complete her undergraduate degree in biology at the University of California. After her prolonged undergraduate studies, Matzinger's PhD took her only three years to complete.

She published her first paper in the venerable *Journal of Experimental Medicine* (JEM) while still a postgraduate student. Lacking confidence, she hoped to impress by adding a co-author – her dog, Galadriel Mirkwood. It was duly published under the authorship

of Matzinger P. and Mirkwood G. But the ruse was discovered, and Polly was barred from the JEM until the offended editor died.

Following post-doctoral studies in Cambridge, England, Matzinger took up a post at the T Cell Tolerance and Memory Section of the Laboratory of Cellular and Molecular Immunology at the National Institute of Allergy and Infectious Diseases. Early on in her tenure her workplace became known as the Ghost Lab because it lay empty for nine months while she investigated the application of chaos theory to the workings of the immune system. She and her colleagues still refer to it as the Ghost Lab on their CVs.

Matzinger concluded in 1994 that for the immune system it was not a question of Self or Non-Self, but of responding to distress signals broadcast by injured cells. In other words, the immune system responds not to attacks but to the damage caused by the attacks. The distress signals are picked up by T cells, which then summon antibodies and macrophages to deal with the damage and the attacker. In the case of tumours, the Danger Model claims, the damage to the cells includes the denial of their ability to send distress signals which would trigger an immune response.

The Danger Model has been refined in the decades since Polly Matzinger proposed it. In an era when deadly new viruses can sweep around the globe in days, immunology has become one of the most important fields in human health studies. Matzinger, now a Section Head at the National Institutes of Health, has been voted one of the fifty most important women working in science today.

# Adam Reiss

## (born 1969)

## Dark energy

In the final decade of the twentieth century it was widely accepted that the universe was created by a single-origin event, the Big Bang, and that it is still expanding because of the energy of that unimaginably powerful explosion: still expanding, but slowing down as the energy is dissipated over the spreading cosmos. Adam Reiss came to a different conclusion.

Our understanding of the cosmos has been transformed in the past hundred years or so with the detection of ever-smaller particles of matter and the science of quantum mechanics, the fundamental processes which make up our universe. Lord Kelvin, in 1884, was one of the first to note that, based on the behaviour of the objects in space that we can see, there must be an awful lot up there that we can't. It's not just invisible planets and stars, which we could detect in conventional ways; the current thinking is that this dark matter is composed of a new as yet unidentified particle. The search for it is now one of the major goals of twenty-first-century astrophysics.

We know from observation and measurement that everything in the universe is getting further apart. The common analogy is with a loaf of raisin bread. As the loaf rises, the perspective of each raisin is that all the other raisins are drifting away from it in the expanding dough. In this metaphor, the energy of the Big Bang is the yeast; and eventually the yeast runs out of energy and the expansion slows and stops.

Astrophysicists know that the universe is still moving outwards because of the change in light frequency and colour, which we on Earth perceive from the most distant objects. Several teams of astronomers in the mid-1990s were making such observations of very distant supernovae, including Adam Reiss at Berkeley in California. By measuring the change in light frequency and the speed of change, they were surprised to find that supernovae billions of years old at the furthest edges of the universe were not slowing down at all. In fact they were accelerating, confounding the law of conservation of energy and the assumptions of decades.

That acceleration must be driven by some form of energy. It cannot be the energy of the Big Bang because that is necessarily spread ever thinner in an ever-expanding universe. Reiss and others – including Brian Schmidt (Reiss's co-researcher at the Mount Stromlo Observatory in Australia) and Saul Perlmutter (who was working on a separate supernova project at Berkeley) – were forced to the conclusion that there must be another source of energy, invisible but powerful.

Both Perlmutter and the Reiss-Schmidt project published their findings in 1998 and the corroborating evidence of the two was convincing. Michael Turner, a Chicago cosmologist and expert in the very earliest moments of the Big Bang, coined the term "dark energy" for this mysterious new force.

There seems no doubt about its existence, but plenty about its nature. Is it simply a property of space, so that as space expands with the residual energy of the Big Bang, more dark energy is automatically created? Did Einstein simply get gravity, which should exist between all objects in space and hold them back, all wrong? Is it yet another undiscovered particle, forming and disappearing with a puff of energy? There are still no answers, and meanwhile astrophysicists are searching for evidence of it in order to build up a body of data from which to draw conclusions.

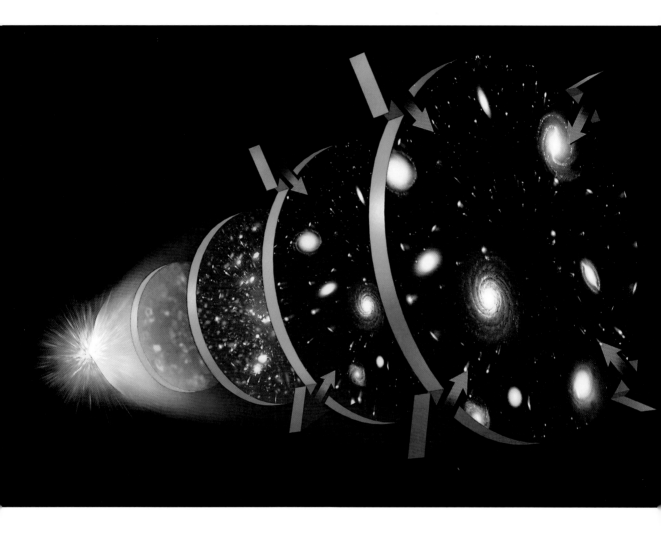

*ABOVE: Illustration of the expansion of the universe. The Cosmos began 13.7 billion years ago in an event dubbed the Big Bang (left). Immediately it began expanding and cooling (stage 1). Eventually, the universe became transparent to radiation, and the first matter was able to form into clumps. Its expansion slowed about ten billion years ago (stage 2). At stage 3, five billion years ago, the universe was full of stars and galaxies, and its expansion began to speed up again because of the mysterious dark energy that pervades the universe.*

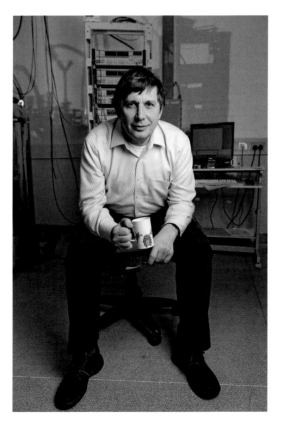

# Andre Geim & Konstantin Novoselov

## (born 1958 / born 1974)

## Graphene

The properties of a new material, graphene, have so excited scientists that they talk of the world entering the Graphene Age. Its formal discovery in 2004 followed centuries of accidental manufacture and decades of theoretical speculation.

Graphene is lightweight, and an extremely efficient conductor of heat and electricity. It is a hundred times stronger than steel. It is almost transparent; yet it absorbs all visible light.

We've probably all made graphene at one time or another. It is a sheet material of pure carbon, only one atom thick. Graphite is composed of many sheets of graphene, and graphite is the material at the core of everyday pencils.

Scientists first began studying the structure of graphite seriously in the early years of the twentieth century; and the structure of layers of graphene within graphite was recognized in 1947 as a factor in graphite's properties of electrical conductivity. German physicist Hanns-Peter Böhm coined the term graphene in 1962 for a single-atom layer of graphite, although he was never able to isolate such a thing.

Efforts to produce graphene began in the 1970s by growing it on other materials, or shaving it from graphite, or simply by dragging it across a surface, like drawing with a pencil. Some of these produced very thin sheets of graphite, but not atom-thin. It fell to a physics professor and a student at the University of Manchester in northern England to isolate graphene in 2004 and properly explore and characterize its behaviour.

Professor Andre Geim and his PhD student Konstantin Novoselov were in the habit of relaxing every Friday evening by, as Geim put it, "messing about in the lab" – setting themselves short experimental tasks unrelated to their usual fields of research. One Friday they decided to investigate graphene, and came up with an embarrassingly easy way of isolating it for the first time.

It's the same method used to remove fluff from woollen sweaters, known in scientific terms as micromechanical cleavage. They pressed a strip of scotch tape against a rough lump of graphite, and pulled it away, bringing a layer of graphene with it – as simple as that. The lump and the tape dispenser are now preserved in the Nobel Museum in Stockholm.

They transferred the graphene from the tape to a strip of silica, which could be used as an electrode when exploring the properties of graphene. The earliest effort was in fact a couple of atoms thick; but within half an hour Geim and Novoselov had made a crude transistor of it. When they succeeded in successively cleaving the layers, they noticed that a one-atom layer of graphene had unique properties quite unlike graphite. "We were very lucky," recalled Geim, "that we had a device with one layer and [another with] two layers of graphene and they behaved completely differently."

When they published their findings, they triggered a frenzy of new scientific research into potential applications. Graphene's properties had implications in fields as diverse as optics and quantum mechanics, chemistry and electromagnetism. Graphene's qualities make it ideal for touch screens and solar cells, and it is expected to replace silicon in solid-state circuitry.

Geim and Novoselov have been characterized as the accidental Nobel winners. But as Geim notes, "Accidents never happen accidently. You actually need to create an environment for those accidents to happen; that's the difference between a good scientist and a bad scientist. So, the good scientists create the environment for as many as possible of those accidents to happen."

# Özlem Türeci and Ugur Sahin

## (2020)

## Coronavirus mRNA vaccine

When the danger of Covid-19 to world health became apparent, the race was on to discover an effective vaccine. Much was at stake, not only in human lives but in the vast sums of financial investment involved in the search, and in the potential profits from success.

Some discoveries are accidents. Some are the result of a deliberate search for something believed to exist. The terrifying spread of Covid-19, the virus first identified in December 2019, focused the collective mind of the international scientific community like no previous emergency.

Coronaviruses are a large family, most of whom cause us little inconvenience beyond a sore throat or a bad cold. They get their name from the rows of protein buds on their surface, whose job it is to identify which cells in the body to attack. If a coronavirus attacks the lungs, it can be a far more serious matter, as the world found out only too well in 2020.

As soon as China released the genetic sequencing of the virus, many pharmaceutical companies accepted the challenge and began to research potential vaccines. They adopted a variety of approaches. Johnson & Johnson based theirs on a rare variant of the coronavirus which causes the common cold. Novavax recreated the Covid-19 protein itself in the laboratory. Astra-Zeneca adapted a coronavirus which normally affects only chimpanzees. It's harmless to humans but carries the necessary genetic code to defeat Covid-19.

Several research efforts tried a new approach to vaccination, using the science of mRNA (messenger ribonucleic acid). The body uses molecules of mRNA to convert DNA into protein. By recreating Covid-19's mRNA, researchers hoped to stimulate the body to produce its own copies of the virus's protein, and then generate an immune response to them.

BioNTech in Germany and Moderna in the US were already conducting research into this technology and took an early lead with their development programmes. By April 2020 BioNTech had a shortlist of four possible vaccines and worked with America's Pfizer Corporation to produce and test them. Pfizer is one of the largest pharmaceutical corporations in the world, founded to produce an antiparasitic drug called santonin in 1849, and BioNTech needed their reach and expertise.

Pfizer identified the most promising one of the four BioNTech variants for further clinical trials in July. Normally a clinical trial at this stage would have involved no more than 300 members of the public. It's an indication of the urgency of the situation that by November 2020 some 40,000 volunteers had received the drug.

Preliminary results suggested that the Pfizer/BioNTech vaccine was 95% effective with very rare serious side effects. There was simply not time to study any long term effects of it. Britain became the first nation to approve the vaccine for emergency use in November, after it had reached the grim milestone of 50,000 deaths due to Covid-19. The United States and the European Union soon followed the United Kingdom's lead; and the Moderna mRNA vaccine was approved for use in January 2021.

Besides giving the world much-needed hope of an end to the pandemic, the success of the two leading mRNA vaccines has wider implications. The technique may offer a new way to combat many other diseases. BioNTech's work is done and it has been suitably rewarded: in the twelve months to February 2021 its share price rose by 156%, making its founders, husband and wife team Dr Ugur Sahin and Dr Özlem Türeci, multibillionaires. Pfizer, which now has the responsibility of manufacturing and distributing the drug, expects to make $15–$30 billion dollars in 2021. The alternative path was taken by British-Swedish firm Astra-Zeneca, who produced the Oxford vaccine for cost. Both have helped make the world a safer place.

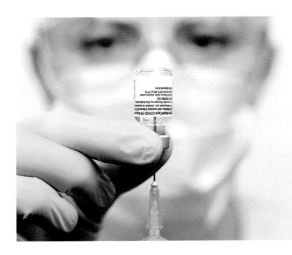

LEFT: *Although the Pfizer BioNTech Covid-19 vaccine needed specialist ultra-cold storage at first, between -60°C and -80°C, subsequent use has shown that this can be reduced to -15°C to -25°C.*

BELOW: *A familiar image from 2020 – a computer-generated graphic of the Covid-19 virus.*

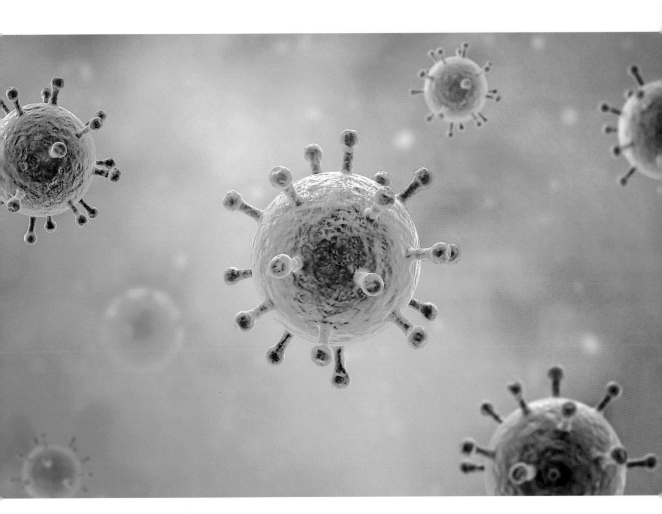

LEFT: *Mars holds the key to future interplanetary discovery. NASA's mission statement is, 'to explore Mars and to provide a continuous flow of scientific information and discovery through a carefully selected series of robotic orbiters, landers and mobile laboratories interconnected by a high-bandwidth Mars/Earth communications network.'*

*ABOVE:* The 'Tianyian' radio telescope in Pingtang County, in southwest China's Guizhou Province. The 500-meter aperture spherical telescope overtook Puerto Rico's Arecibo Observatory, which was previously the world's largest at 300 meters in diameter. Arecibo, on the island of Puerto Rico, provided the first reliable data on gravitational waves.

# Index

*ABOVE: Danish astronomer (or astrologer as they were then known) Tycho Brahe achieved astonishingly accurate results without the use of a telescope.*

*ABOVE: Danish physicist Niels Bohr who helped establish quantum theory and for whom the element bohrium is named.*

*ABOVE: Marie Curie is one of only three scientists to win two Nobel prizes for their scientific work. Daughter Irène became a Nobel Laureate in 1935 and both her children became prominent French scientists.*

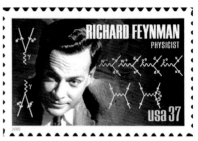

*ABOVE: Influential American physicist Richard Feynman with a stamp illustrating his Feynman Diagrams.*

*ABOVE: Albert Einstein photographed on a visit to Washington, D.C. in 1921, the year he won the Nobel Prize for Physics.*

*ABOVE: An image of the Sombrero Galaxy (Messier 104) taken with NASA's Hubble Space Telescope in 2003.*

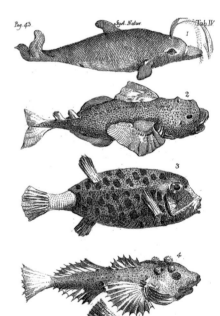

*ABOVE: An illustration of bony fish from a later edition of Carl Linnaeus's* Systema Naturae.

*BELOW: A tunnel view of the Large Hadron Collider (LHC) which spans the border of France and Switzerland.*

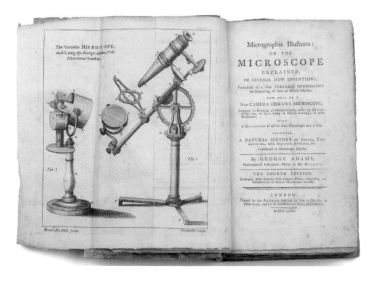

ABOVE: *The success of Robert Hooke's* Micrographia *spawned a series of books illustrating the microscopic world, this one by instrument maker George Adams of Fleet Street, London.*

*ABOVE: Science was not immune to the political cartoonist. This one by James Gillray, in 1802, features
the popular science lectures at the Royal Institution. Two politicians to the left are engaged with a gaseous
experiment, while behind them, holding the bellows, is Sir Humphry Davy, pioneer of nitrous oxide.*

# Acknowledgements

Colin Salter is grateful to Donald King, who taught him both science and history at Lathallan School in Montrose, Scotland. With this book on the history of scientific discovery he repays, in part, the debt he owes Mr King, who made both subjects fun to learn. The seeds of fascination which Mr King planted have flourished. Colin has also written the *Science is Beautiful* series.

**Also in this series:**

**Maps That Changed the World** ISBN 978-1-84994-297-3
John O. E. Clark (2015)
**100 Diagrams That Changed the World** ISBN: 978-1-84994-076-4
Scott Christianson (2014)
**100 Documents That Changed the World** ISBN: 978-1-84994-300-0
Scott Christianson (2015)
**100 Books That Changed the World** ISBN: 978-1-84994-451-9
Scott Christianson and Colin Salter (2018)
**100 Speeches That Roused the World** ISBN: 978-1-84994-492-2
Colin Salter (2019)
**100 Letters That Changed the World** ISBN: 978-1-911641-09-4
Colin Salter (2019)
**100 Children's Books That Inspire Our World** ISBN: 978-1-911641-08-7
Colin Salter (2020)
**100 Posters That Changed the World** ISBN: 978-1-911641-45-2
Colin Salter (2020)

**About the Author**

Colin Salter is a history writer with degrees from Manchester Metropolitan University, England and Queen Margaret University in Edinburgh, Scotland. His most recent publications are *The Moon Landings: One Giant Leap*, *Science is Beautiful: The Human Body Under The Microscope*, *Science is Beautiful: Disease and Medicine*, *100 Books That Changed the World*, *100 Speeches That Roused the World* and *100 Letters That Changed the World*. He is currently working on a memoir based in part on letters written to and by his ancestors over a period of two hundred years.

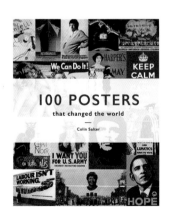